THE RISE OF SUPERCONDUCTORS: RACE TO THE FUTURE OF TECHNOLOGY

Written by:
Austin Mardon,Kendall Caperchione, Sudipta Samadder,
Alicia Au, Mariyam Sardar, Janani Rajendra, Maria Gonzalez,
Darla Chloe Daniva, Aleefa Devji, Lajendon Jeyakumar,
Gabriela Ivanov, Sifar Halani

Edited by:
Kanish Baskaran

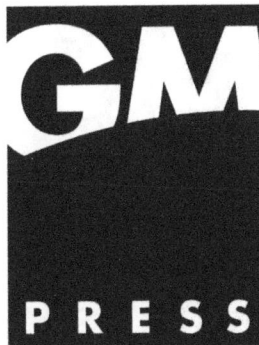

GM PRESS

Typeset and Cover Design by Kim Huynh

ISBN 978-1-77369-593-8
Golden Meteorite Press
103 11919 82 St NW
Edmonton, AB T5B 2W3
www.goldenmeteoritepress.com

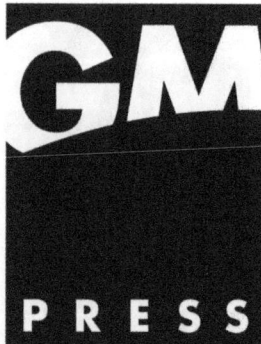

Table of Contents

Table of Contents

Table of Contents

Table of Contents

Introduction

Superconductors are materials capable of achieving zero electrical resistance and expulsion of magnetic fields. They have been described as the "future of technology" and have a wide variety of applications from transportation to quantum computing. This book delves into the topic of Superconductors, including its history and discovery, applications, controversy and impact on society.

1. What are Superconductors?

By Kendall Caperchione

Introduction

Electricity is one of the foundational marks of modern society, as most modern technologies hinge on the concept of electrical output and how it is conducted. Medicine, transportation, communication, entertainment, and infrastructure are just a few of the sectors of everyday life that are heavily impacted by the conduction of electricity through connectors and conductors. Consider your home. The conduction of electricity powers your bedside lamp, pushes energy into your laptop battery, provides heat in the winter, and air conditioning in the summer. Electrical flow pushes electricity through the wire conductors within your home to provide a multitude of tasks and actions to allow you to go about your daily business and routine. However, what about the other side of electricity? How about the side of natural conduction where elements and physical substances are able to conduct electricity through themselves? Or the idea of superconductivity? These are all different types of electrical conduction that we do not necessarily think about on a day-to-day basis, as it is less frequent in occurrences in repetitive routines.

Electrical conduction is a natural phenomena, and this chapter aims to talk specifically about a specific applicator of conduction; superconductors. Superconductors, as will later be explained in this chapter, are considered any sort of element or "metallic alloy" that has a major decrease in electrical resistance when cooled to a certain temperature or temperature threshold (Jones, 2019). These types of natural elements and metals are able to conduct electricity through them without loss of energy or speed, which is often considered a marvel by modern scientists as it is usually very difficult to identify and create a natural superconductor (Jones, 2019).

To further understand the history of superconductors, how they are produced, or merely the science involved in the process of conduction, it is important to have a grasp on what specifically superconductors are, which is exactly what this chapter will do. By dividing specific topics into two main sections, this chapter will break down the basics of superconductors in an easy to understand way, to ensure that the background knowledge you will gain will further allow you to understand the rest of the topics within this book. To start, the first half of the chapter will attempt to explain and define what superconductors are. "What is a super conductor?", "What are some of the properties of superconductors?", "What are the applications of superconductors?", and "How do we identify what elements are superconductors?" are amongst some of the major questions this section will tackle in hopes of laying the basic foundation for the second half of the chapter. Similar to the first section, the second section will also be broken down into four (4) different subsections, where the classifications of superconductors will be explained. Some of these include TypeI/II superconductors, different elemental properties of superconductors, and the main difference between a semiconductor and superconductor. This section will also analyze why different conductors can be categorized into unique classification sectors, based on their inherent characteristics. In the hopes of offering a simplistic understanding for this complex topic, the best place to begin and to understand the idea of natural superconductors is through answering the burning question of *what are superconductors anyway?*

What are Superconductors?

How Do We Define What A Superconductor Is?

As mentioned, superconductors are elements and natural metallic materials that provide an unobstructed transition and movement of an electrical current when the temperature reaches a cool temperature threshold (Jones, 2019). Superconductors in essence provide an electrical current to pass without the loss of energy, in turn creating a type of supercurrent that conducts electricity through the element or substance (Jones, 2019). The threshold can be represented by the 'critical temperature', which is often unmeasurable due to the rareness and uniqueness of the low rate at which the temperature must be for the conduction to work (*Critical Temperature*, n.d.).

Typically, metals also present some resistance when at room temperature ratings, however when these metals reach the critical temperature point (which is different for every element and metal), the electrons are able to bounce from atoms with easy in a turn creating what is known as superconductivity, specifically through metals (*Superconductors: Types, Materials, & Properties*, 2020). During the 20th century, some of the major testing done to metals and elements was used in order to test the conductivity of said material, working from materials such as lead, to other elements and physical materials such as carbon nanotubes (*Superconductors: Types, Materials, & Properties*, 2020). Superconductors are used across various fields, particularly in medicine with the magnetic resonance imaging (MRI) machine, where electrons flow freely through the conductor after the critical temperature is met.

The superconductor materials however are not easy to find, and because of this, materials with low energy status must be used to conduct these experiments. Although there are a few identifiable superconductors right now, there are still studies and academic research being completed in order to find and develop compounds that can be turned into superconductors after reaching that critical temperature point (*Superconductors: Types, Materials, & Properties*, 2020).

Properties of Superconductors

Superconductors are often the elements and compounds that show not only similar attributes and characteristics, but are the ones that uniquely react to the critical temperature for their own element. Some of the major properties include their ability to conduct electricity for an infinite amount of time, the Meissner Effect, and persistent currents to name a few (*Superconductors: Types, Materials, & Properties*, 2020). The idea of infinite conductivity explains how a superconducting material presents no resistance when processing electrons, after the cooling of the element to its critical temperature point (Madhav University, n.d.).

One superconductor that shows this trait is Mercury, as it creates no resistance for atoms at 4k (Kelvins) (Madhav University, n.d.). The second defining property of superconductors is a theory called the Meissner Effect, which is also known as when a superconductor is able to expand the magnetic field around an atom (Madhav University, n.d.). This theory, pioneered by German physicist W. Meissner in 1933 is used in the 21st century to explain the process through which an element becomes a superconductor. In this process, the magnetic field is cooled to the critical temperature which is when the atom will expel the magnetic field, further creating a fluid area for electrons to freely transition and flow between atoms (*Meissner Effect*, n.d.).

The Meissner Effect is also effectively used in determining different types of superconductors, as the test is able to determine which elements and metals are either wholly or partially able to conduct electrons with minimal resistance after reaching the critical temperature (*Meissner Effect*, n.d.). The final property that is commonly associated with superconductors is the persistent current that flows through the material. When a superconductor ring is present within the magnetic field, the superconductor can reach the critical temperature threshold, and when this ring is abandoned, the flow through the current becomes persistent and generates and influx of electrons that bounce in the ring (*Superconductors: Types, Materials, & Properties*, 2020). The most important piece of information to remember about the persistence of currents is how the consequence is always that the current will flow with no resistance within the continuous loop of material (OpenLearn, n.d.).

Applications of Superconductors

Notably, not all materials are able to complete superconductivity. In fact, most of the elements that are able to effectively carry out superconductivity are alloys and metals, which make up approximately 25 elements known on the periodic table (Woodford, 2021). An important idea to consider is the application of superconductive materials in everyday life and everyday use. For example, a power plant that uses electricity and superconductivity to reach your home is using it in a more effective way than regular conduction (Woodford, 2021).

Electrical applications of superconductors have been used for decades, as it creates a more efficient way of transporting and utilizing electricity. From the use of coal and fossil fuels beginning in the 19th century, this only accounts for approximately 20% of current energy sources that power items and actions used everyday, as most companies and players within the energy sector are using the transmission of superconductors to transport electricity through generators and other electrical devices (Woodford, 2021). Not only has super conduction created a space within the energy sector of clean and efficient transition of energy on large scales, but this concept has also created a more effective way of implementing conduction to smaller electronic devices, like computers and household devices (Woodford, 2021).

Superconductors are also responsible for the impact on magnetic devices and applications. Magnetic effects are usually seen within the medical field in body scanners, specifically in MRI's, where the atoms give off radio waves in order for the electrons to engage in low temperature superconductivity (Woodford, 2021). Magnetic applications have also been seen in other physics based fields of science where CERN has implemented practices of superconductivity for their LHC, or more well known Large Hadron Collider, in which superconductivity is used in particle acceleration as the magnetic fields within the curved area smash together in a deeper way creating more effective results (Woodford, 2021).

How to Identify Superconductors

So far we have established that superconductors are identified based on specific properties that can be tested, and with certain results provided the identity can be confirmed. Mercury has been included in the list of superconductors as not only is it the most popular superconductor, but it is also the first to be discovered all the way back in 1911 (Jones, 2019).

After a study emerging out of the University of Cambridge was able to provide insight into the induction of superconductors in science and society, there was also a unique set of circumstances outlined that pertain directly to the identification of superconductors. The major discovery was that superconductor elements had charge density waves present, which creates the effect of rippled electrons that are twisted to create free-moving electrons between atoms (University of Cambridge, 2014).

Many superconductors were not discovered by scientists until the late 20th century, as most of the superconductor elements and compounds required a cooling temperature of 0 degrees (on the Kelvin scale), or the equivalent of -273 degrees celsius (University of Cambridge, 2014). High temperature superconductors however remain an issue with scientists and physicists as low temperature conductors are easier to identify, and high temperature conductors remain reluctant to deduce properties or common variables (University of Cambridge, 2014).

Classifications of Superconductors

Type-I/Type-II Superconductors

Since there are plenty of elements and metals that can be identified as superconductors, and of those materials, only some can conduct perfectly, scientists and physicists have categorized superconductors into different sections based on characteristics.

Beginning with the major classifications of superconductors, there are two main types of conductors; Type-I and Type-II. The first type, Type-I superconductors, are materials that show some conductivity when the critical temperature is between 0 and 10 kelvins (Superconductors and Superconducting Materials Information, n.d.). These materials are typically metals and metalloids, and usually reach superconductivity through the regression in speed causing cooler and lower temperatures (Superconductors and Superconducting Materials Information, n.d.). Type-I superconductors, also known as 'soft' conductors, are often the materials and metals that remain in a superconducting state for a shorter amount of time as their weak nature allows for the process to halt (Akshit, 2021). As mentioned, mercury is one of the most common Type-I superconductors, however some of the other Type-Is include aluminium, lead, and other compounded elements like silicon carbide with boron (Akshit, 2021).

The second sector of superconductors is Type-II conductors, or also referred to as 'hard' conductors. Type-II superconductors usually conduct electricity at a higher temperature than the soft conductors, as they will conduct at up to 130 kelvins resistance free. Type-IIs are also a lot more versatile than Type-Is, as they are able to carry larger amounts of electrical current without losing their superconductivity, as a result of the stronger magnetic fields that retain the Type-II conduction properties (Akshit, 2021).

Semiconductor/Superconductors

Superconductors can also be distinguished from their counterpart, semiconductors. To first review what superconductors are, there are some distinct properties and observations that we can consider. Firstly, superconductors are the types of elements and compounds that offer zero amounts of electrical resistance when conducting electricity. This process ensures that electrons that pass through the course lead to a loss of resistance as the energy gap of superconductors is a lot higher than semiconductors (Superconductors: Types, Materials, & Properties, 2020). Superconductors are unique to the elements and compounds discussed in modern science as they are able to facilitate a conductive rate higher than regular conductors. As previously stated, superconductors allow for electrical transfer without the loss of energy, and with the critical temperature at their own specific threshold are the unique superconductor materials (Madhu, 2020).

The other side of the comparison offers a different type of conductor, and one that is not as intense as superconductors. Semiconductors are conductors that have 'finite resistivity' and have an energy gap smaller than superconductors (Superconductors: Types, Materials, & Properties, 2020). One of the other defining aspects of a semiconductor is in the way that they lie in the values between insulators and conductors, which makes them reactive to certain activities such as light and magnetic fields (Madhu, 2020). Some metals which have been mentioned already in this chapter including silicon, tellurium, and tin, are all considered semiconductors. Silicon itself is arguably the most mainstream semiconductor as it is used in the production of wired circuits in the electrical industry (Madhu, 2020).

Metals

Metals are unique elements on the periodic table that allow for conductivity to occur when the circumstances allow for it. When the charged particles reach the critical temperature, the electrons are able to move through the element, allowing for electricity to flow freely through the metal (Gorski, 2019). Two of the most common metals that are used by the electrical industry for energy conduction are copper and silver. Both of these metals are often the most common metal conductors as they have what is considered to be a higher amount of 'movable atoms' which allow for the electrons to pass more freely without resistance, in turn meaning the more motion of the atoms within the metal, the better the conductivity (Gorski, 2019). Silver, unlike copper, is more expensive to produce and is typically used for specialized equipment or particular manufacturing purposes. For example, silver is more commonly used in the construction of satellite boards and circuits but may not be used in everyday households as T.V. connectors or charger cords (Gorski, 2019).

Another metal that is commonly used as a cheaper conductor is aluminum. Aluminum is usually the more common material, similar to copper, as it is generally found in household objects. However, aluminum is a concerning metal to use for general consumption as the metal allows for a build up of an oxide surface that will create electrical resistance for currents passing through (Gorski, 2019). This can become a major problem as the likelihood of an overheating cord or circuit becomes more probable, and in turn is typically used for physically larger conductors like transmission lines on hydro towers or other high-voltage terrains where it is safer to use for professionals and not everyday people (Gorski, 2019).

Despite these three metals listed, there are a multitude of other elements that can be used for conduction including (but not limited to) gold, steel, brass, and iron.

Insulators

This final subsection of the chapter aims to analyze the opposite of what conductors do, and those elements or materials that in essence stop the natural flow of electrons. Insulators are the items that do just that. Focusing on protection, insulators are primarily used in homes and electrical circuits to prevent the flow of electricity when the heat becomes too much and could cause damage or dangerous conditions (Ring, 2019). Since insulators are seen as the opposite of a conductor, the same is true for their properties. Insulators have electrons that stay on the atom, or are static, and with this, the atoms and electrons that try to flow through the conductor are stopped due to the resistance, or are concentrated to a singular section to create a more efficient conductor (Ring, 2019).

Conclusion

Superconductors are an interesting feat of the electrical industry, as they provide electrical currents and electricity to many of the everyday devices and appliances that people use in their homes, vehicles, work, and for leisure. This chapter aimed to give the reader a basic overview of what superconductors are and how they work. The first section of the chapter looked at what superconductors are, and did an in-depth review of how to identify them, their properties, and their applications. The second section of the chapter further took into consideration the classifications of superconductors, specifically examining Type-I and Type-II, semi and superconductors, metals, and insulators.

The basic information provided in this chapter will give you the pre-knowledge needed for the rest of the book, which will take a further analysis into multiple different ideas surrounding the superconductor. Some of the further topics include the history and discovery of superconductors, the science/physics behind superconductivity, the manufacturing of the product in commercial uses, and future uses of superconductors, to name a few. In closing, superconductors and conductivity is the basic foundation of electrical transportation, which runs our society today, and with it, our lives have been drastically improved and have become more efficient because of these tools that scientists, physicists, and other electrical professionals implement everyday.

References

Akshit. (2021, June 16). Superconductors: Types & Examples. StudiousGuy. https://studiousguy.com/superconductors-types-examples/#Type-I_Superconductors.

Encyclopædia Britannica, inc. (n.d.). Critical Temperature. Encyclopædia Britannica. https://www.britannica.com/science/critical-temperature.

Encyclopædia Britannica, inc. (n.d.). Meissner effect. Encyclopædia Britannica. https://www.britannica.com/science/Meissner-effect.

Gorski, V. (2019, March 2). What Metals Make Good Conductors of Electricity? Sciencing. https://sciencing.com/metals-make-good-conductors-electricity-8115694.html.

Jones, A. Z. (2019, May). What Is a Superconductor? Definition and Uses. ThoughtCo. https://www.thoughtco.com/superconductor-2699012.

Madhu. (2020, May 26). Difference Between Semiconductor and Superconductor. Compare the Difference Between Similar Terms. https://www.differencebetween.com/difference-between-semiconductor-and-superconductor/.

Ring, J. (2019, March 2). What Are Insulators? Sciencing. https://sciencing.com/insulators-8031301.html.

Superconductors and Superconducting Materials Information. Superconductors and Superconducting Materials Selection Guide | Engineering360. (n.d.). https://www.globalspec.com/learnmore/materials_chemicals_adhesives/electrical_optical_specialty_materials/superconductors_superconducting_materials.

Superconductivity. http://www.madhavuniversity.edu.in/. (n.d.). https://madhavuniversity.edu.in/superconductivity.html.

Superconductivity. OpenLearn. (n.d.). https://www.open.edu/openlearn/science-maths-technology/engineering-technology/superconductivity/content-section-2.2.

Superconductor : Types, Materials, Properties and Its Applications.

ElProCus- Electronics | Projects | Focus. (2020, February 3). https://www.elprocus.com/what-is-superconductor-types-materials-properties/.

University of Cambridge. (2014, June 17). Superconducting secrets solved after 30 years. Phys.org. https://phys.org/news/2014-06-superconducting-secrets-years.html.

Woodford, C. (2021, March 11). How do superconductors work? Explain that Stuff. https://www.explainthatstuff.com/superconductors.html.

2. History/background of Superconductors

By Sudipta Samadder

Introduction

Superconductivity is one of nature's most intriguing quantum phenomena. In the 1800s, scientists could explain what occurred in superconductivity but were unable to solve the why and the how of superconductivity for nearly 50 years. For instance, before discovering superconductors, it was already known that cooling a metal increased its conductivity. This was due to decreased electron–phonon interactions. Over the years, there had been many significant discoveries that built up the theory of superconductivity and led to the identification of materials that were superconductive. This chapter aims to explore the history and background of superconductivity, including the events that led up to its discovery.

Exploring ultra-cold phenomena (to 1908)

The history of superconductivity dates all the way back to the second half of the nineteenth century, when scientists were devising methods to condense gases with low boiling points. This is because superconductors require super low temperatures to function. James Dewar was a British chemist and physicist who invented a double-walled vacuum flask in 1982 that was very effective in insulating cold liquids (Encyclopedia Britannica, 2021). In particular, he initiated research into electrical resistance at low temperatures. Zygmunt Florenty Wróblewski, a Polish physicist and chemist, later researched the electrical properties at low temperatures, though his research ended early due to his accidental death. (Chem Europe, n.d). While he worked as an assistant under Professor Jolly, he conducted a series of experiments and wrote a dissertation named 'The Investigation of the Induction of Electric Energy by Mechanical Means' (Renan et al., n.d). Later in his life, he introduced a modern chemical laboratory with the latest electrical appliances, which was one of the first electrical installations in Krakow (Renan et al., n.d). Around 1864, scientists Karol Olszewski and Wróblewski predicted the electrical phenomena in ultra-cold temperatures of low resistance levels. This was documented by Olszewski and Wróblewski in the 1880s (Chem Europe, n.d). Later in 1885, L.P. Cailletet (1832-1913), E. Bouty (1846-1922) and Z.F. Wroblewski (1845-1888) conducted the first systematic studies of the dependence of electrical resistance on temperature. Their research led them to conclude that it would be reasonable to expect a zero value for the resistance for a temperature higher than -273 degrees celsius (Gavroglu, 2009). The next set of extensive measurements of the electrical resistance of various metals were performed by James Dewar (1842-1923) and John Ambrose Fleming (1849-1945). One prediction that Dewar and Fleming made was that pure metals would become perfect electromagnetic conductors at absolute zero temperature (Chem Europe, n.d). In 1896, they completed a study of the resistance of mercury at liquid air temperature. From their observations, they concluded that the resistance of mercury could vanish at zero degrees (Gavroglu, 2009). Later on, Walther Hermann Nernst developed the third law of thermodynamics and stated that it was impossible to attain the temperature of absolute zero in any real experiment (Schummer, 2018). Moreover, the commercial researchers, Carl von Linde and William Hampson, filed for patents on the Joule-Thomson effect nearly at the same time. Linde's patent involved the use of regenerative counterflow method to produce a systematic investigation of established facts (Chem Europe, n.d). Hampson also used designs based on a regenerative method. The combined process of Hampson's and Linde's design became known as the Linde-Hampson liquefaction process. This process involves the liquefaction of gases and introduces a regenerative cooling positive feedback system. The Linde machine was used by scientist Heike Kamerlingh Onnes for his own research (Chem Europe, n.d). On March 21, 1900, a U.S patent was granted to Nikola Tesla for the purpose of increasing the intensity of electrical oscillations by lowering the temperature (Chem Europe, n.d). This phenomenon was previously observed by Olszewski and Wroblewski. This patent describes the increased intensity and duration of electric oscillations of a low-temperature resonating circuit (Chem Europe, n.d). It is believed that Tesla had intended to use Linde's machine to attain cooling agents (Chem Europe, n.d). To conclude, the researchers involved in investigation of the ultra cold phenomena provided clues that led to better understanding of superconductivity.

Sudden and fundamental disappearance

The first observation of superconductivity was reported in 1911, by Dutch physicist Heike Kamerlingh Onnes of Leiden University. This observation was made with mercury. Since then, many other superconducting materials have been discovered, ultimately building up the theory of superconductivity. One of the fundamental investigations that contributed to this discovery was the reinvestigation of Dewar's earlier experiments on the reduction of resistance at low temperatures. This reinvestigation was led by Onnes and Jacob Clay. Onnes and his assistants began the investigations on platinum and gold, replacing these later with mercury since it is a more readily refinable material. Onnes accomplished his research on the resistivity of solid mercury at cryogenic temperatures by using his own process of attaining liquid helium as a refrigerant (Chem Europe, n.d). He found that at 3 K, the value of resistance of pure mercury became 0.0001 times the value of the resistance of solid mercury at 0 degrees celsius (Gavroglu, 2009). Later that year, as he investigated with temperatures reaching 4.19 K, he observed that the resistivity abruptly disappeared. This meant that the measuring device Onnes was using did not indicate any resistance (Chem Europe, n.d). In 1911, Onnes wrote a research paper titled "On the Sudden Rate at Which the Resistance of Mercury Disappears". In the paper, he specifically stated that relative to the best conductor at ordinary temperature, specific resistance" becomes thousands of times less. A significant discovery to note is that when Onnes later revised the process, he found that at 4.2 K, the resistance returned to the material. In the following year, Onnes published more articles about the phenomenon. Upon upon his discovery, Onnes initially called the phenomenon "supraconductivity" (1913) but later changed the term to "superconductivity". For his research, Onnes was awarded the Nobel Prize in Physics. The research did not end there. To test the application and usability of superconductivity, in 1912, Onnes conducted an experiment where he introduced electrical oscillations into a conductive ring and took out the battery that generated electrical oscillations. When he measured the electric current of his system, Onnes found that the intensity of electrical oscillations did not go down. Ultimately, this provided experimental proof of Tesla's patent. The reason why the current's lifespan was increased was because of the superconductive state of the conductive medium. In the following decades, superconductivity was discovered in several other materials. For example, in 1913, lead was found to superconduct at 7 K, and in 1941, niobium nitride was found to superconduct at 16 K (Chem Europe, n.d). These discoveries were significant in producing high performance electric power transmission systems.

Enigmas and solutions

The next important step in understanding superconductivity occurred in 1933. This is when Meissner and Ochsenfeld discovered that superconductors expelled applied magnetic fields, a discovery which has come to be known as the Meissner effect. Later in 1935, F. and H. London showed that the Meissner effect was caused by a superconducting current which carried electromagnetic free energy that had been minimized. In 1950, Landau and Ginzburg devised the phenomenological Ginzburg-Landau theory of superconductivity. The Ginzburg-Landau theory, which combined Landau's theory of second-order phase transitions with Schrodinger-like wave equation, was very successful in explaining the macroscopic properties of superconductors. These properties were used to classify superconductors into two categories now referred to as type I and type II. This discovery helped Abrikosov and Ginzburg win the 2003 Nobel Prize. An additional discovery was made in 1950 by Maxwell and Reynolds et al. They found that the critical temperature of a superconductor depends on the isotopic mass of the constituent element. This important discovery helped to explain the microscopic mechanism responsible for superconductivity which is the electron-phonon interaction (Chem Europe, n.d).

By 1957, the complete microscopic theory of superconductivity was finally proposed by John Bardeen, Leon N. Cooper, and Robert Schrieffer. It was called The BCS theory. The BCS theory explained the superconducting current as a superfluid of Cooper pairs, which are pairs of electrons interacting through the exchange of phonons. In honour of this work, the authors were awarded the Nobel Prize in Physics in 1972. The BCS theory became very prominent and widely recognized in 1958, when Nikolay Bogolyubov showed that the BCS wavefunction could be obtained using a canonical transformation of the electronic Hamiltonian.

In the following year, Lev Gor'kov demonstrated that the BCS theory breaks down to the Ginzburg-Landau theory close to the critical temperature. Gor'kov was the first person to derive the superconducting phase evolution equation.

Another significant discovery associated with the BCS theory is the little -parks effect. The little-parks effect was discovered in 1962 when experiments were conducted with empty and thin-walled superconducting cylinders exposed to a parallel magnetic field. The electrical resistance of such cylinders shows a periodic oscillation with the magnetic flux through the cylinder. William Little and Ronald Parks provided an explanation that the resistance oscillation reflects the periodic oscillation of the superconducting critical temperature (Tc). Tc refers to the temperature at which the sample becomes superconducting. The Little-Parks effect is caused by the collective quantum behaviour of superconducting electrons.

Superconductivity in the 1980s

Discoveries in the field of superconductivity was unrivaled in the in the years leading up to the 1980s. In 1964, a research by the name of Bill Little from Stanford University had suggested the possibility of organic (carbon-based) superconductors (Ankara University, n.d). The successful synthesis of these types of superconductors was performed in 1980 by Danish researcher Klaus Bechgaard of the University of Copenhagen along with his three French team members. (TMTSF)2PF6 is an organic conductor that had to be cooled to a very cold transition temperature of 1.2K, and subjected to high pressure to superconduct (Ankara University, n.d). The mere existence of this organic conductor proved the possibility of "designer" molecules. These are molecules modified to function in a predictable way. Following this discovery, a flurry of activities in the field of superconductivity had been triggered (Ankara University, n.d). Around the world, researchers began experimenting and subsequently producing very imaginable combinations of substances in order to search for higher and higher Tc's (Ankara University, n.d). In January of 1987, a team of researchers from the University of Alabama-Huntsville replaced Yttrium for Lanthanum in the Muller and Bednorz molecule and achieved a significant value of 92K Tc. For the first time, a material which is referred to as YBCO had been found to superconduct at temperatures warmer than liquid nitrogen (a commonly available coolant) (Ankara University, n.d). In addition to YBCO, many exotic - and often toxic - elements in the base perovskite ceramic have contributed to significant milestones. Specifically, the mercuruc-cuprates are the current class (or "system") of ceramic superconductors (Ankara University, n.d). It was in 1993 when the first synthesis of one of these compounds were achieved at the University of Colorado. It was synthesized by A. Schilling, M. Cantoni, J.D. Guo, and H.R. Ott of Zurich, Switzerland (Ankara University, n.d). Moreover, the Tc of 138 K is now held by a thallium-doped, mercuric-cuprate composed of the elements Mercury, Thallium, Barium, Calcium, copper and oxygen. In February of 1994, the Tc of this ceramic superconductor was verified by Dr. Ron Goldfarb at the National Institute of Standards and Technology-Colorado. (Ankara University, n.d). It is worth noting that under extreme pressure, its Tc can be pushed up even higher - approximately 25 to 30 degrees more at 300,000 atmospheres (Ankara University, n.d).

There are several companies that capitalize on high-temperature superconductors. The first company was Illinois Superconductor which was formed in 1989, and is known as ISCO International today (Ankara University, n.d). The company introduced a depth sensor for medical equipment that was able to operate at liquid nitrogen temperatures (~77K). In 1997, researchers discovered that an alloy of gold and indium was both a superconductor and a natural magnrt at a temperature very close to absolute zero (Ankara University, n.d). In theory, it was believed that a material with such properties could not exist (Ankara University, n.d). However, since then, over a half-dozen of such compounds have been discovered. In recent years, researchers have also discovered the first high-temperature superconductor that does not contain any copper. The first all-metal perovskite superconductor was also found (Ankara University, n.d). Furthermore, in 2001, a material that had been stored on laboratory shelves for decades was found to be a significantly new and well capable superconductor (Ankara University, n.d). Additionally, a group of Japanese researchers measured the transition temperature of magnesium diboride $(MgB2)$ at 39 kelvin, which is far above the highest Tc of any of the elemental or binary alloy superconductors (Ankara University, n.d). While 39 K is still well below the Tc's of the "warm" ceramic superconductors, subsequent modifications in the way $MgB2$ is made have opened opportunities for its use in industrial applications (Ankara University, n.d). Through laboratory testing, it was found that $MgB2$ will outperform NbTi and Nb3Sn wires in high magnetic field applications like MRI (medical resonance imaging) (Ankara University, n.d). There is no doubt that the applications of superconductors are numerous. The research into superconductivity is ongoing. What's more is that the groundbreaking discoveries that are currently being made are continuously advancing our understanding and applications of superconductivity.

Conclusion

The discovery of superconductors and the theory of superconductivity has been groundbreaking in modern science. In the early years of superconductivity, progress to applications was slow and intermittent. Following the discovery of superconductors by a team of scientists in the late nineteenth century to the early twentieth century, superconductors were found to create powerful magnets for devices, such as MRI. By

1980, superconductivity had been observed in many metals and alloys. In summary, this chapter shows that the history of superconductivity has been full of surprises and that superconductivity is a stimulating and continuing problem in physics.

References

Ankara University. (n.d). History of Superconductors. http://cesur.en.ankara.edu.tr/history-of-superconductors/

Chem Europe. (n.d). History of Superconductivity. https://www.chemeurope.com/en/encyclopedia/History_of_superconductivity.html

Encyclopedia Britannica. (2021). Sir James Dewar. https://www.britannica.com/biography/James-Dewar

Gavroglu, K. (2009). Superconductivity. Compendium of Quantum Physics. 750-757. https://link.springer.com/chapter/10.1007/978-3-540-70626-7_215

Renan, Springer, Zeeman. (n.d). Zygmunt Florenty Antonovich Wróblewski. Prabook. https://prabook.com/web/zygmunt_florenty.wroblewski/721075

Schummer, J. (2018). Walther Nernst. Encyclopedia. https://www.encyclopedia.com/people/science-and-technology/chemistry-biographies/walther-hermann-nernst

3. Discovery of Superconductors

By Alicia Au

Introduction

In this chapter, the origin of superconductors will be discussed. The discovery of superconductors by Heike Kamerlingh Onnes along with the efforts of various individuals in the field will be recounted and analyzed to showcase the team effort that is needed in science advancement. These stories highlight both their serendipitous and non-serendipitous scientific discoveries and their role in advancing the future of science, particularly in the field of physics.

The Pioneer: Heike Kamerlingh Onnes

Born on September 21, 1853, in Groningen, Heike Kamerlingh Onnes studied physics and mathematics at the University ofGroningen in 1870 (Reif-Acherman, 2013). He demonstrated great strength in academia as he passed the test for a bachelor's degree a year later (Reif-Acherman, 2013). He then studied under notable scientists in another university but returned to Groningen University in 1873 to receive his doctoral degree at the level of magna cum laude in 1879 (Reif-Acherman, 2013). He then continued in academia as an assistant at a Polytechnic School in Delft until 1882 where he was later appointed Professor of Experimental Physics and Meteorology and Director of the Physics Lab at the University of Leiden for the next forty-two years (Reif-Acherman, 2013). It is impressive to note that he was only twenty-nine years old at that time (Reif-Acherman, 2013). During his time as the Director of the Physics lab, he showed interest in the behavior of matter at low temperatures from the beginning of his assignment (Reif-Acherman, 2013). As a result, he started an experimental research program with devices that had the ability to accurately measure volume, pressure, and temperature for the matter that was in liquid and gaseous states (Reif-Acherman, 2013). The accuracy of the devices was of the utmost importance for him as he believed that it was crucial to finding the underlying patterns that were needed for the study (Reif-Acherman, 2013). This was reflected in his motto that is posted on his physics labs that read "through measurement to knowledge-Door Meten tot Weten" (Reif-Acherman, 2013).

The Key Experiments

In the boom of the science field in the late 1800s and throughout the 1900s, empiricist thought prevailed, leading to races and competitions amongst scholars and researchers to find solid evidence that would verify hypotheses in experimental practices. Given the time, these races were limited by access to both physical resources and monetary constraints, and so creativity and ingenuity were key to making valid discoveries. The race to discover superconductivity was not exempt from this. The first kind of permanent gaseous oxygen to liquid form was made by L. Cailletet in France half a decade before Kamerlingh Onnes (Dahl, 1984). In 1883, Wroblewski and Olszewski first produced liquid oxygen that boiled quietly in a test tube (Dahl, 1984). Olszewski later was on a mission to liquefy hydrogen which meant he was joining a race between Kamerlingh Onnes and Dewar (Dahl, 1984). As a turn of events, Dewar won the discovery of liquefying hydrogen in 1898 and solidifying hydrogen a year later (Dahl, 1984). That left only one permanent gas left to be liquified which was helium. This race between Olszewski, Dewar, and Onnes continued and William Ramsay joined later on; however, all of the contenders were at a disadvantage because they lacked the helium gas that Onnes had access to (Dahl, 1984).

Kamerlingh Onnes's introduction to this field of work was largely inspired by his colleague and friend Johannes Diderik van der Waals who is famous for two theories: an equation of state for real gases in 1873 and his principle of corresponding states which was an extension of his equation which allowed it to be written as a function of critical pressure, volume, and temperature in 1880 (Reif-Acherman, 2013). Kamerlingh Onnes wanted to verify Van der Waals' theories and further their applications in low temperatures (Reif-Acherman, 2013). He decided to do this by doing experiments on the liquefaction of gases.

In the first couple of years of his experiment, Kamerlingh Onnes devoted his time to finding the right equipment for his lab to conduct accurate experiments at low temperatures (Reif-Acherman, 2013). In addition, in preparation, he studied the works of many physicists around the world such as Martinus van Marum who studied the liquefaction of ammonia, Charles Saint-Ange Thilorier who studied carbon dioxide, Micheal Faraday who studied chlorine, and Charles Cagniard de la Tour and Thomas Andrews work on the critical point (Reif-Acherman, 2013). Moreover, he reviewed the work of liquefaction of other elements that had been done at the time such as oxygen by Louis Paul Cailletet, oxygen and nitrogen by Zygmunt Wróblewski and Karol Olszewski, and of hydrogen by James Dewar (Reif-Acherman, 2013). Even with all the ample preparation of studies on the liquefaction of gases, Kamerlingh Onnes's experiment was interrupted for two years starting in 1896 because of the hazard of working with large amounts of compressed gases (Reif-Acherman, 2013). This nevertheless, pushed him to improve his strategy and instrumentation in his lab by both acquiring and constructing new instruments. He focused so much on instrumentation that he supported the later making of the Leiden School for Instrument Makers for trade workers that were constructing his complicated laboratory instruments (Reif-Acherman, 2013). Kamerlingh Onnes is actually known for his style of research as it displays "meticulous planning, precise measurement and constant improvement of techniques and instruments' ' (Reif-Acherman, 2004).

By 1892, Kamerlingh Onnes had built air liquefiers with improved processes compared to Pictet's that were available at the time (Reif-Acherman, 2013). Many milestones followed in the years following such as obtaining liquid oxygen, obtaining a bath of liquid oxygen measuring -193 degrees celsius (Reif-Acherman, 2013). In 1898, the Dutch Supreme Court finally gave permission for his laboratory to continue under restrictions (Reif-Acherman, 2013). This allowed for him to liquefy hydrogen reaching temperatures of -250 degrees celsius (Reif-Acherman, 2013). Every couple of years, he would upgrade the instruments in his labs which would lead to being able to produce 71 liters of hydrogen in 1906. This allowed him to use cryostat baths of liquid hydrogen sustainably (Reif-Acherman, 2013).
On the tenth of July 1908, Heike Kamerlingh Onnes was able to produce over 60 cubic meters of liquid helium for the first time (Reif-Acherman, 2013). This experiment that lasted for fourteen hours gained him recognition in the natural science field (Reif-Acherman, 2013). At that moment, Kamerlingh Onnes' laboratory for a brief moment was the coldest place on Earth (Delft, 2008). The result of this experiment opened many doors in the field of low-temperature physics, a field that was in its infancy at the time. This field would only become more and more complex

and impactful as time progressed, leading to many modern scientific revolutions, particularly in the field of cryogenics. Onnes' liquefaction of helium has been recognized by many to be the most important day of Kamerlingh Onnes's career after the fact. This is why he was awarded a Nobel prize in physics for that discovery in 1909 (Reif-Acherman, 2013). Kamerlingh Onnes had gotten three more Nobel prize nominations in his lifetime, another Nobel prize in 1912, and one more in 1913 (Reif-Acherman, 2013). With his Nobel prize from 1909, he attracted many notable scientists such as Marie Curie and Albert Perrier to his lab (Reif-Acherman, 2013). Marie Curie later worked with him on radium radiation at liquid hydrogen temperatures which won her a Nobel prize in chemistry in 1911 (Reif-Acherman, 2013).

Kamerlingh Onnes's laboratory assistant Clause August Crommelin also worked with him on the experiment (Dahl, 1984). This partnership led to Kamerlingh Onnes analysis of the temperature dependence of electrical conductivity (Dahl, 1984). They continued the work of Dewar and John Ambrose Fleming whose experiments showed that the electrical resistance of pure metals was minuscule close to absolute zero (Dahl, 1984). Kamerlingh Onnes carried out experiments to further investigate this area of research which led to the discovery of superconductivity in April 1911 (Dahl, 1984). He was studying the resistance of solid mercury at cryogenic or super cold temperatures using liquid helium which he was able to produce in his laboratory because of his past discovery (Dahl, 1984). During the experiment, Mercury was cooled to approximately four degrees Kelvin, a temperature just above absolute zero, just below the boiling point of helium when it was observed that the electrical resistance suddenly disappeared (Dahl, 1984). As mentioned in the what are superconductors chapter, it was named a superconductor because there is a small energy loss due to resistance (Dahl, 1984). This led to five Nobel prize nominations in 1912 (Dahl, 1984). From here, Kamerlingh Onnes gained interest in the possibility of using superconductors to produce high field electromagnets (Dahl, 1984).

The Discovery

The Nobel prize is awarded to individuals based on three criteria according to historian Elisabeth Crawford: "(1) that the works rewarded should constitute "discoveries," "inventions," or "improvements"; (2) that these should represent "the most modern results" and that "works or inventions of older standing ... be taken into consideration only in case their importance has not previously been demonstrated"; (3) that these should be works that "have conferred the greatest benefit on mankind" (Reif-Acherman, 2013).

The Dutch experimental physicist, Kamerlingh Onnes was then awarded the Physics Nobel prize in 1913 for his work and contribution to low-temperature physics. The experiment that was credited for the award was the production of liquid helium in 1908. The success of the production of liquid helium experiment birthed the field of cryogenics and the discovery of superconductivity a couple of years later in 1911 (Reif-Acherman, 2013).

The story of the now century-old discovery that carved new paths in physics has been revisited and has been found to have been misrecorded for many decades. Heike Kamerlingh Onnes documented his serendipitous discovery in his lab notebooks in great detail at the time of the discovery of superconductivity; however, it was reported to be difficult to access three years later (Van Delft & Kes, 2010). This lack of information led to inaccurate recounts of events or even inaccurate accreditation of the discovery of superconductors to circulate. One example of misinformation that was widely accepted as fact was the role of a sleepy apprentice in his lab (Van Delft & Kes, 2010). This tale was even published in a 1996 Physics Today article by Jacobus De Nobel. In addition to this, there were also rumors of Kamerlingh Onnes's laboratory notebooks disappearing. This called for a revisit to Kamerlingh's notebooks by Van Delft and Kes in 2010 through the Kamerlingh Onnes Archive at the Boerhaave Museum in Leiden, where Kamerlingh Onnes' laboratory was and continues to be situated.

Future additions to the discovery of superconductors

Although Kamerlingh Onnes is credited with the discovery of superconductors, advances and applications of his theory have been done by his precedents in his fields such as John Bardeen, Leon Cooper, Robert Schiffer, Fritz Walther Meissner, and Robert Ochenfeld. Just as the discovery of superconductors has created a new field of research in physics, the preceding discoveries in edits are recognized to have the same impact in the field of physics as well.

John Bardeen, Leon Cooper and Robert Schriffer,

The Bardeen-cooper-Schrieffer theory, also known as the BCS theory of superconductivity, was created by John Bardeen, Leon Cooper, and

Robert Schrieffer. This theory explains the "why" behind Onnes' fingsings. Why do the electrons flow without resistance in certain materials at low temperatures? Bardeen, Cooper, and Schrieffer theorize that it is because "a single negatively charged electron slightly distorts the lattice of atoms in the superconductor, drawing toward it a small excess of positive charge" which "attracts a second electron. It is in this weak, indirect attraction that binds the electrons together into a cooper pair" (Joint Quantum Institute, n.d.). This theory won the group a Nobel prize in physics in 1972.

Fritz Walther Meissner and Robert Ochsenfeld

The German pair of physicists, Fritz Walther Meissner and Robert Ochsenfeld discovered the Meissner effect, also called the Meissner-Ochsenfeld effect in 1933 (Libretexts, 2020). It describes the phenomenon of expelling the magnetic field from within that happens with superconductors (Libretexts, 2020). This only happens when the superconductor is below its critical temperature meaning that the material has to have zero resistance and therefore exhibit superconductor properties. This temperature was previously found to be different for different materials due to their bond length and strength. The Meissner effect allows magnets to levitate above a superconductor which is an important property for its applications that will be discussed in the chapter discussing the magnetic applications of superconductors.

Conclusion

There were many key factors that played a role in the discovery of superconductors. It was a combination of Kamerlingh Onnes's emphasis on having the right instruments that provided the possibility of quality quantitative measurements in low temperatures along with the opportunities that he had with his academic connections with well-established scientists. Moreover, the friendships and drive that came with the influence of notable scientists that were friends, colleagues, or mentors of his such as J.D. van der Waals which allowed Kamerlingh Onnes to be so successful in making the discoveries that lead to the discoveries of superconductors. In addition, in examining the discovery story of the superconductors, healthy competition in the field of science can be examined. The competition between Kamerlingh Onnes, Olszewski, Dewar, and William Ramsay allowed for the acceleration of the progress in science. The origin story of natural scientific concepts can allow for the appreciation of the world in its intricacies. The human race has been pursuing explanation questions of the surroundings to have a greater purpose and understanding of the world. This has allowed humanity to civilize and use the knowledge that has been acquired to innovate, create and continue to ask questions. In hopes of recounting the story of not only one Nobel prize winner but the group effort of all the scientists that have allowed this discovery to flourish, collaboration within fields, between fields, and outside of academia is encouraged.

In the next chapter, the manufacturing process of superconductors will be discussed. The method used to acquire superconductors in a laboratory setting in the 1900s will be drastically different in the mass consuming habits of the mass that the industry has to satisfy. Kamerlingh Onnes' contribution to the research about superconductors is not only the application of the matter at that temperature but also the apparatuses needed for the manufacturing that is still needed today. The application of both his contributions as well as the collaboration and contributions of other scientists, engineers, designers will be discussed in the next chapter about manufacturing.

References

Dahl, P. F. (1984). Kamerlingh onnes and the discovery of superconductivity: The leyden years, 1911-1914. Historical Studies in the Physical Sciences, 15(1), 1-37.

Joint Quantum Institute. (n.d.). Bardeen-Cooper-Schrieffer (BCS) Theory of Superconductivity. Joint Quantum Institute. https://jqi.umd.edu/glossary/bardeen-cooper-schrieffer-bcs-theory-superconductivity.

Libretexts. (2020, July 19). Meissner Effect. Engineering LibreTexts. https://eng.libretexts.org/Bookshelves/Materials_Science/Supplemental_Modules_(Materials_Science)/Magnetic_Properties/Meissner_Effect.

Reif-Acherman, S. (2013). Heike Kamerlingh Onnes and the Nobel Prize in Physics for 1913: The Highest Honor for the Lowest Temperatures. Physics in Perspective, 15(4), 415-450.

Van Delft, D. (2008). Little cup of helium, big science. Physics today, 61(3), 36.

Van Delft, D., & Kes, P. (2010). The discovery of superconductivity. Physics Today, 63(9), 38-43.

Van Delft, D. (2012). History and significance of the discovery of superconductivity by Kamerlingh Onnes in 1911. Physica C: Superconductivity, 479, 30-35.

4. Manufacturing of superconductors

By Mariyam Sardar

Introduction

Superconductors have two beneficial properties. At a critical temperature, they conduct electricity without resistance (Ndahi, 2003). In addition, they also have magnetic properties (Ndahi, 2003). Some conductors are manufactured as a means to conduct electricity, whereas other superconductors are manufactured as magnets. There are two requirements for superconductors to be deemed good; they must be inexpensive to manufacture, and they must be easy to manufacture (Bottura & Godeke, 2012). The materials for superconductors are selected based on cost, availability, material properties, and environmental conditions (Ndahi, 2003). There are various methods to manufacture a superconductor. To overcome a compound's physical property, such as brittleness, new techniques have to be considered. Some of the processes discussed in this chapter are the top-seed melt texture growth (TSMTG) process, sol-gel method, oxide-powder-in-tube (OPIT) process, and various shock techniques. In addition, this chapter addresses specific procedures of manufacturing some low-temperature superconductors (Nb3Sn and Bi-2212) and high-temperature superconductors (GdBCO and ceramics). Due to the complex nature of the manufacturing process, there is a likely chance that the superconducting phase is not achieved. Hence, a lot of the time, superconductors are unsuccessful. In industrial places, these failed superconductors are discarded. To use the material of the failed superconductors, a study has found a technique to recycle the material from the failed superconductors (Shi et al., 2015).

Processes Used to Manufacture Superconductors

Top-seeded Melt Texture Growth (TSMTG) Process

The TSTMG process requires placing a structurally alike seed crystal of similar lattice parameters "on the top surface of a mixed pre-synthesis precursor powder pellet prior to melting process" (Wang et al., 2018). The above orientation helps to "control...the spontaneous nucleation and growth of the grains" (Wang et al., 2018). The arrangement is then heated at ~1060oC, where the materials turn into a molten state (Wang et al., 2018). Subsequently, the sample is heated at 1045oC for peritectic solidification (Wang et al., 2018). Next, the sample is cooled down, where a peritectic reaction nucleates at the seed (Wang et al., 2018). To be precise, a non-superconducting phase reacts with the liquid phase to form the superconducting phase (Wang et al., 2018). The traditional TSMTG process requires synthesizing the pure compound phase before producing superconducting materials (Wang et al., 2018). For example, to produce the Gd2BaCuO5 superconductor via the traditional TSMTG process, the synthesis of Gd123 and GD211 is required (Wang et al., 2018).

Sol-Gel Method

In the Sol-Gel Method, the powders of metallic compounds and solvents are mixed (Aşikuzun & Öztürk, 2020). Then, the prepared mixtures are stirred with a heated stirrer in a closed environment until a clear solution is obtained (Aşikuzun & Öztürk, 2020). Succeedingly, the stirring process is continued until the solution is transformed into a gel-like texture (Aşikuzun & Öztürk, 2020). A study by Aşikuzun and Öztürk used the sol-gel method for manufacturing the YBa2Cu3-xZnxO superconductor. First, yttrium (III) acetate hydrate, barium acetate, copper (II) acetate, and zinc acetate dihydrate powders were mixed with acetic acid and methanol (Aşikuzun & Öztürk, 2020). The solution was stirred for 8 hours with a heated magnetic stirrer to obtain a clear solution (Aşikuzun & Öztürk, 2020). Next, the solution was stirred for an additional 12 hours to convert the solution into a gel-like structure (Aşikuzun & Öztürk, 2020).

Oxide-Powder-In-Tube Process (OPIT)

In the OPIT process, the loose powder is filled in a tube and is compressed to form a bar (Seifi et al., 2000). The compacted bar is placed in a tube (Siefi et al., 2000). Following, the tube is annealed to form a thin wire (Seifi et al., 2000). Multiple single-filaments are arranged in a new tube, where the multiple-filament tube is also annealed to form a wire (Seifi et al., 2000). The multi-filament wire is pressed to form thin bendable tapes (Seifi et al., 2000). The tapes are heated, which converts the materials to superconducting materials (Seifi et al., 2000).

Various Shock Techniques for the Consolidation of Powders

Shock processes are used to compact or sinter metallic powders (Mammalis, 2001). Some shock processes are shock-compaction, shock-enhanced sintering, shock-conditioning, shock-induced chemical synthesis, shock-induced phase transformations, chemically assisted shock consolidation (Mammalis, 2001). Shock-compaction is the solidification of the powders through the application of shock energy at particular surfaces (Mammalis, 2001). Shock-enhanced sintering is when the solidification of the powders is compressed and heated in the sintering procedure to reach the final product (Mammalis, 2001). Shock-conditioning is a process where the powder is shocked, remilled and sintered (Mammalis, 2001). Shock-induced chemical synthesis is a technique where the powdered compound is formed and solidified simultaneously (Mammalis, 2001). Shock-induced phase transformation is when the desired compound is formed under high pressure (Mammalis, 2001). Chemically assisted shock consolidation is a combination of two processes: shock-induced chemical synthesis and shock consolidation (Mammalis, 2001). The above processes are applied in the various machinery to attain the final product of the powders.

Manufacturing of Low-temperature Superconductors (LTS)

Nb3Sn manufacture

In 1961, the Nb3Sn superconductor was made by filling crushed powder of niobium and tin into a niobium-plated wire (Bottura & Godeke, 2012). NB3Sn was one of the first high-field superconductors (Bottura & Godeke, 2012). After the formation of Nb3Sn, in order for Nb3Sn wires to act as a superconductor, the temperature must be less than 18 K (Barzi and Zlobin, 2019). The industrial Nb3Sn superconductors are manufactured in three ways: bronze route, internal tin, and powder-in-tube (Bottura & Godeke, 2012).

The bronze-route wire is composed of Nb/Nb-alloy filaments that are assembled in a Sn-rich bronze mould (Bottura & Godeke, 2012). The compound (Nb-Sn bronze) is inserted in a can of Cu stabilizer, and a thin diffusion barrier is placed between the bronze and Copper (Bottura

& Godeke, 2012). The diffusion barrier is an inert of Cu, such as Nb, Tb, or Va (Bottura & Godeke, 2012). To compact to the final wire diameter, the wire is heated to the temperature range of 600oC to 700oC (Bottura & Godeke, 2012). The presence of Cu destabilizes the Sn-rich line compounds (Bottura & Godeke, 2012). The cooling down prevents excessive grain growth, which increases the pinning efficiency (Bottura & Godeke, 2012). The tin diffuses in the Cu-Sn matrix, and then the Sn reacts with niobium to form the superconductor, Nb3Sn (Bottura & Godeke, 2012). The major disadvantage of the bronze route is the low solubility of Sn in the bronze matrix (Bottura & Godeke, 2012). As a result, stoichiometry is not achieved (Bottura & Godeke, 2012). Also, the production of Nb3Sn using the bronze route is not superconducting at high magnetic fields (Bottura & Godeke, 2012). The max density that superconduction of Nb3Sn works for this method is 1000 A/mm2 at 4.2 K and 12 T (Bottura & Godeke, 2012).

Another method that fixes the major disadvantage of the bronze route is the internal tin technique (Bottura & Godeke, 2012). As suggested by the name, internal tin technique, Sn and Nb filaments are placed in the Cu matrix, which is surrounded by a barrier to prevent any undesirable reactions (Bottura & Godeke, 2012). Furthermore, multiple filaments of Cu matrixes are enclosed in a pure copper can (Bottura & Godeke, 2012). Later on, the filaments are heated, and the Sn and Cu matrix diffusion occurs (Bottura & Godeke, 2012). The diffusion results in the Sn reacting with Nb to form Nb3Sn (Bottura & Godeke, 2012).

The last method to manufacture Nb3Sn is the powder-in-tube process (Bottura & Godeke, 2012). The stacking tubes of Nb are filled with Nb2Sn powder and some Cu additive (required for destabilization of Sn) (Bottura & Godeke, 2012). The stacking tubes are placed into the Cu matrix (Bottura & Godeke, 2012). Following this, the Cu matrix is heated to form the end product, Nb3Sn (Bottura & Godeke, 2012).

However, some of the challenges with the production of Nb3Sn are as follows. The production of Nb3Sn conductors was not widely adopted due to issues with the element's characteristics, such as material brittleness and long filament (Bottura & Godeke, 2012). Also, to date, the manufacturing of Nb3Sn is costly (Bottura & Godeke, 2012). Hence, the manufacturing of Nb3Sn is not extensively practiced (Bottura & Godeke, 2012). The manufacturing of Nb3Sn is limited to 2 tons per year (Bottura & Godeke, 2012).

The Bi-2212 advantage is that it is available as a round wire and can form into a Rutherford cable (Bottura & Godeke, 2012). However, the disadvantage is the low magnetic field. The critical density of round wire is 500 A/mm2 at 20 T and 4.2 K (Bottura & Godeke, 2012). The manufacturing of Bi-2212 is similar to Nb3Sn due to the brittleness of the Bi-2212 (Bottura & Godeke, 2012). The process is known as the Wind-and-React magnet fabrication process (Bottura & Godeke, 2012). The manufacturing of Bi-2212 requires a 100% oxygen atmosphere and 890oC temperature (Bottura & Godeke, 2012). Until recently, the Bi-2212 was only manufactured for the purpose of high field magnets because the blockage of current flow was unknown (Bottura & Godeke, 2012). It was recently found that the blockage was due to bubbles that formed inside the Bi-2212 (Bottura & Godeke, 2012). There are two reasons why bubbles formed (Bottura & Godeke, 2012). First, in the composition of Bi-2122, 25% of hollow space is required in the matrix for the hardening of Bi-2122 (Bottura & Godeke, 2012). The extra space results in the accumulation of bubbles during the melting reaction (Bottura & Godeke, 2012). The second reason is that during the melting of Bi-2122, the rapid release of oxygen from the powder amplifies the bubble formation (Bottura & Godeke, 2012). In addition, other contaminants, such as carbon and hydrogen, can react with oxygen to form carbon dioxide and water (Bottura & Godeke, 2012). However, the process to remove the bubble is still unknown, and hence, more research is needed to figure out the current flow of Bi-2122.

Manufacturing of High-Temperature Superconductors (HTS)

GdBCO

The GdBCO superconductors can be manufactured from raw materials using the TSTMG method (Wang et al., 2018). Compared to the traditional TSTMG method, the GdBCO method is less expensive (Wang et al., 2018). The cost of raw materials - Gd_2O_3, BaO, and CuO oxides - is cheaper than the synthesis of Gd123 and Gd211 (Wang et al., 2018). The precursor powder was prepared from raw materials (Wang et al., 2018). The powder is pressed into the cylindrical pellet, and a support layer is also pressed into the plate (Wang et al., 2018). The seed, NdBCO, was positioned at the top surface of the precursor pellet (Wang et al., 2018). The pellets were placed on an alumina plate with some MgO crystals (Wang et al., 2018). The samples were heated to 900oC and secured for 20 hours

(Wang et al., 2018). Then, the sample was heated up to 1060oC and was held for 2 hours to guarantee that the decomposition of the Gd123 phase was complete (Wang et al., 2018). Later, it was cooled down via the two-step sintering process (Wang et al., 2018). The sample was cooled down to 1035oC, followingly cooled down to 1015oC and then cooled down to room temperature (Wang et al., 2018). Lastly, in the presence of oxygen, the sample was heated at a temperature between 330°C – 410°C for 200 hours (Wang et al., 2018). Now, the melting of the raw materials would have superconducting properties (Wang et al., 2018). Wang and others called this TS-MOMG, the top-seeded metallic oxide melt growth.

Ceramics

Manufacturing ceramic superconductors is a modern technique (Seifi et al., 2000). The OPIT process and explosive compaction technique are utilized to assemble the ceramic superconductors (Seifi et al., 2000; Mammalis, 2001). It uses shock waves to compress the ceramic powder containing lead, calcium, bismuth, strontium, and copper oxides to form solid components (Seifi et al., 2000; Mammalis, 2001). The shock waves are passed through the explosive detonation to the porous media, which creates high shock pressures, resulting in the breakdown of the original compounds and sintering (Mammalis, 2001). The compact ceramic material is inserted into a silver tube and is heated to form a silver wire that contains ceramic materials (Seifi et al., 2000). The previous single-silver-filament wires are arranged to make a multi-filament wire in a Ag tube (Seifi et al., 2000). Similar to the annealing of the single-filament wire, the silver tube of the multi-filament is also annealed to form a wire (Seifi et al., 2000). The wires are rolled into lengthy tapes (Seifi et al., 2000; Mammalis, 2001) or extruded to form rods or wires (Mammalis, 2001), or other specific machines can be used to manufacture precise geometric superconductors (Mammalis, 2001). After the shape of the ceramic-silver filament is confirmed, the wire is heated to 835oC (Seifi et al., 2000). The heat treatment results in diffusion and grain growth, leading to ceramic-superconducting properties (Seifi et al., 2000; Mammalis, 2001).

Recycling Failed Superconductors

The rare earth materials such as Sm, Gd and Y, and the Ba-Cu-O are high-temperature superconductors that produce high magnetic fields (Shi et al., 2015). However, the cost of these raw materials and the low yield of the superconductors has been an issue in manufacturing

REBCO superconductors (Shi et al., 2015). The TSMG process is used to manufacture the REBCO superconductors (Shi et al., 2015). Many dopants -Y-211, CeO2, Pt, 2411(M), Ag, and ZrO2 - are researched to enhance the properties of the REBCO superconductors (Shi et al., 2015). However, all these dopants make the growth process intricate, and many adjustments are needed to fabricate the REBCO superconductors (Shi et al., 2015). The TSTMG process is complicated (Shi et al., 2015). As the growth phase is vulnerable to changes, any changes to the growth phase result in the risk of failure of grain growth (Shi et al., 2015). Usually, the raw materials are discarded due to no use of the mixed raw materials (Shi et al., 2015). However, Shi and others have found a method to recycle the failed superconductors into raw materials.

The REBCO sample was recycled by replenishing the liquid phase lost during the growth process without grinding the failed REBCO into powder (Shi et al., 2015). The REBCO (without Ag2O) mixed powder was pressed and placed on the top of the failed sample (Shi et al., 2015). The REBCO was placed to help the seeding process and to avoid any contamination (Shi et al., 2015). Below the failed sample, the Yb pressed pellet was placed to support the shape of the liquid phase (Shi et al., 2015). Succeeding, the sample was placed in the box furnace for the melting process (Shi et al., 2015). The samples were heated to 1015oC, then cooled down to 1008oC, 985oC and lastly to room temperature (Shi et al., 2015). The samples were oxygenated for ten days at 450oC - 420oC (Shi et al., 2015). The end product was the fully grown recycled samples on the top surface (Shi et al., 2015). Using this method, they were able to yield 90% of the raw materials from the failed superconductors (Shi et al., 2015).

Conclusion

The superconductors are essential in current times and are very efficient due to no electrical resistance. The superconductors are manufactured in various methods due to the different properties of materials. Some of the currently practiced processes to manufacture the superconductors are the TSTMG process, sol-gel method, OPIT process, and various shock techniques. In addition, these processes are applied in manufacturing the LTS (such as Nb3Sn and Bi-2122) and HTS (such as GdBCO and ceramics). Also, at times, the manufacturing of the superconductors is unsuccessful; hence, a new method is introduced to recycle the failed superconductors. As a matter of fact, new materials and new methods are constantly researched to reduce the cost while maintaining an easy manufacturing procedure.

References

Aşikuzun, E., & Öztürk, Ö. (2020). Theoretical and Experimental Comparison of Micro-hardness and Bulk Modulus of Orthorhombic YBa2Cu3-xZnxO Superconductor Nanoparticles Manufactured using Sol-Gel Method. Sakarya University Journal of Science, 24(5), 854–864. https://doi.org/10.16984/saufenbilder.676028

Barzi, E., & Zlobin, A. V. (2016). Research and Development of Nb3Sn Wires and Cables for High-Field Accelerator Magnets. IEEE Transactions on Nuclear Science, 63(2b), 783–803. https://doi.org/10.1109/TNS.2015.2500440

Bottura, L., & Godeke, A. (2012). Superconducting Materials and Conductors: Fabrication and Limiting Parameters. Reviews of Accelerator Science and Technology, 5(1), 25–50. htt ps://doi.org/10.1142/S1793626812300022

Mamalis, A. G. (2001). Near net-shape manufacturing of bulk ceramic high- Tc superconductors for application in electricity and transportation. Journal of Materials Processing Tech, 108(2), 126–140. https://doi.org/10.1016/S0924-0136(00)00741-X

Ndahi, H. B. (2003). Manufacturing with Superconductors. Technology Teacher, 63(3), 17–20. ht tps://link.gale.com/apps/doc/A111304349/AONE?u=anon~2bf66719&sid=googleScholar&xid=0638f09a

Seifi, B., Bech, J. I., Eriksen, M., Skov-Hansen, P., Wang, W. G., & Bay, N. (2000). Manufacturing of Superconducting Silver/Ceramic Composites. CIRP Annals - Manufacturing Technology, 49(1), 185–189. https://doi.org/10.1016/S0007-8506(07)629 25-4

Shi, Y., Namburi, D. K., Wang, M., Durrell, J., Dennis, A., & Cardwell, D. (2015). A Reliable Method for Recycling (RE)-Ba-Cu-O (RE: Sm, Gd, Y) Bulk Superconductors. Journal of the American Ceramic Society, 98(9), 2760–2766. https://doi.org/10.1111/jace.13683

Wang, M., Wang, X., Hu, C., Ma, J., & Yang, W. (2018). A Low-Cost Fabrication Technique for the Growth of Single-Domain GdBCO Bulk Superconductor from Raw Metal Oxides. Journal of Superconductivity & Novel Magnetism, 31(12), 3835–3840. https://doi.org/10. 1007/s10948-018-4673-0

5. Magnetic applications of superconductors

By Janani Rajendra

Superconducting Magnets

Superconducting magnets are a form of electromagnets that are made from coils of superconducting wire and must be cooled to cryogenic temperatures while they are in use (Zinkel, 2020). At very low temperatures, superconductivity causes materials to lose their electrical resistance which prevents the superconductor from heating up, allowing them to conduct very strong electrical currents (The magic of superconductors in the spotlight, n.d.). These magnets have the ability to create very strong magnetic fields because when they are in the superconducting state, the wire near the magnet has no electrical resistance, allowing the magnet to produce significantly greater electric currents compared to the average electromagnet (Zinkel, 2020). Although superconducting magnets are impressive and valuable to society, these magnets can be quite costly to use, due to the amount of energy lost as heat in the windings of the wire (Zinkel, 2020). However, as time has passed, more effective methods have been discovered. Later on in this chapter, a more efficient material used in superconductors will be discussed.

Superconducting magnets often contain an immense amount (hundreds of litres) of liquid helium (Hockings et al., 2010). The energy is gathered in the superconducting coils of the magnet and eventually empties into the cryogenic liquid (Hockings et al., 2010). An abundant amount of cryogen gas is distributed into the surrounding environment very quickly since the expansion factor for liquid helium is 760:1. Even though manufacturers design magnets with safety in mind, there still remains a risk of asphyxiation, a state of lacking oxygen which can potentially be lethal, due to the opaque fog of helium and nitrogen gas that takes over the oxygen in the room where the magnet is located (Hockings et al., 2010). To prevent adverse outcomes, such as the one described in the previous sentence, rooms with these magnets are required to have an oxygen detector, in order to detect when the oxygen levels in the room are below what is considered safe (Hockings et al., 2010). Further, rooms with these magnets also have a quench vent to ensure that the released gases are able to exit the room efficiently (Hockings et al., 2010).

Large Hadron Collider

The Large Hadron Collider is globally considered to be one of the biggest applications of superconductivity, and it contains 23 kilometers of superconducting magnets surrounding its 27 kilometer circumference (The magic of superconductors in the spotlight, n.d.). It currently operates using energy of about 6.5 TeV per beam (Pulling together: Superconducting electromagnets, n.d.). This energy allows trillions of particles to circle the collider's 27 kilometer tunnel 11,245 times per second (Pulling together: Superconducting electromagnets, n.d.). The Large Hadron Collider features a series of interconnected linear and circular accelerators which speed up the particles to their maximum speed (Pulling together: Superconducting electromagnets, n.d.). This process requires more than 50 types of magnets in order to send the particles through the sophisticated paths without the particles decreasing in speed because without any force, the particles would eventually drift apart and travel in a straight path (Pulling together: Superconducting electromagnets, n.d.). There is a circular accelerator in this instrument which allows the magnetic field to keep the particles in their orbits; however, higher energy particles need a much stronger field compared to lower energy particles (The magic of superconductors in the spotlight, n.d.). Thus, the energy of circular accelerators, such as the Large Hadron Collider, are restricted by the power of their magnets (The magic of superconductors in the spotlight, n.d.). This limitation posed a major issue at the end of the 1960s because it restricted the amount of progress that could be made (The magic of superconductors in the spotlight, n.d.). However, with the discovery of superconductivity it was possible to overcome this challenge (The magic of superconductors in the spotlight, n.d.).

The first superconducting collider, a project known as Tevatron, was started in 1983 (The magic of superconductors in the spotlight, n.d.). This project successfully drove the use of superconductors for high energy physics (The magic of superconductors in the spotlight, n.d.). Later on, the Large Hadron Collider was invented, after which followed other superconductors which now influence accelerator projects, such as High Luminosity LHC (The magic of superconductors in the spotlight, n.d.). In the future, it is expected that larger colliders will break even more boundaries of the currently existing energy levels, allowing for further exploration and discovery (The magic of superconductors in the spotlight, n.d.).

Nuclear Magnetic Resonance (NMR)

Nuclear Magnetic Resonance, also known as NMR, is a physical phenomenon that has many applications in the field of physical sciences (Kaseman & Iyer, 2020). NMR is a nuclear specific spectroscopy that investigates the underlying spin characteristics of atomic nuclei by utilizing a large magnet (Kaseman & Iyer, 2020). Similar to other types of spectroscopies, NMR also uses electromagnetic radiation (radio frequency waves) to enhance the progressions between various nuclear energy levels—this capability is known as resonance (Kaseman & Iyer, 2020). The discovery of NMR has been quite useful to scientists as this technology has empowered scientists to determine the structure of small molecules (Kaseman & Iyer, 2020). In addition, NMR is also pivotal in the study of genomics, drug discovery, biotechnology, and various other material sciences (Coalition for the Commercial Application of Superconductors, 2014).

The nucleus of an atom is made up of elementary particles, neutrons and protons, which have an intrinsic property known as spin (Kaseman & Iyer, 2020). Quantum numbers can be used to describe spin and atomic nuclei (Kaseman & Iyer, 2020). An even number of protons and neutrons are said to have zero spin, while atomic nuclei with an odd number of protons and neutrons are said to have a non-zero spin (Kaseman & Iyer, 2020). As such, molecules with a non-zero spin have a magnetic moment (Kaseman & Iyer, 2020). The magnetic moment causes the nucleus to behave as miniature bar magnets (Kaseman & Iyer, 2020). Without the presence of an external magnetic field, each magnet is oriented in random orientations in space, whereas when there is an external magnetic field, such as during an NMR experiment, the bar magnets are forced to orient themselves with the low energy or against the high energy (Kaseman & Iyer, 2020).

The applications of NMR are critical in the fields of medicine and chemistry, without regard to the many applications being discovered on a daily basis (Kaseman & Iyer, 2020). NMR provides high quality images of the body in areas of cardiovascular, neurological, musculoskeletal, and oncological imaging (Kaseman & Iyer, 2020). Compared to alternative methods, such as computed tomography (CT), NMR does not use ionized radiation which makes this method much safer (Kaseman & Iyer, 2020). Laboratories use NMR to determine structures of chemical and biological compounds through various peaks given by the NMR spectra (Kaseman & Iyer, 2020). Moreover, NMR is also very useful in other fields such as in environmental testing, the petroleum industry, process control, in Earth's field NMR, and in magnetometers (Kaseman & Iyer, 2020). Since NMR is non-invasive, it is a very cost-efficient method for testing expensive biological samples and allows for multiple experimental trials if necessary (Kaseman & Iyer, 2020).

Superconducting Magnets in NMR

Superconducting magnets is the core foundation of the NMR spectrometer (Coalition for the Commercial Application of Superconductors, 2014). Low temperature superconductor materials play a vital role in enabling the homogenous magnets of NMR spectroscopy to operate at acute precision (Coalition for the Commercial Application of Superconductors, 2014). Increased field values as well as increased homogeneity and stability, achieved by the superconducting magnets, is what enables the magnetic field to produce top tier resolution needed for the determination of protein structure and other NMR analysis (Coalition for the Commercial Application of Superconductors, 2014).

The ground-breaking discoveries do not stop here; over the past 25 years, advancements have been made to the superconducting materials (Coalition for the Commercial Application of Superconductors, 2014). Niobium-titanium was the first conductor used, after which niobium-tin conductors were used, and now a vast assortment of varying materials are able to be used in ultra-high field applications (Coalition for the Commercial Application of Superconductors, 2014). These leading-edge improvements lead to the development of even higher field NMR spectrometers that have the capability to analyze more sophisticated molecules (Coalition for the Commercial Application of Superconductors, 2014).

Granting all this, there still remains limitations on the availability of superconducting materials used in NMR, says a report from the National Academies of Science (Coalition for the Commercial Application of Superconductors, 2014). The current superconducting materials are expected to limit prospective NMR machines to the 1000 MHz (1 GHz) level; however, scientists would like to eventually exceed this level (Coalition for the Commercial Application of Superconductors, 2014). Advancements in the development of higher field superconductors are recommended to further improve the existing NMR technology and to provide more cutting-edge applications from the newly developed NMR technology (Coalition for the Commercial Application of Superconductors, 2014).

Magnetic Resonance Imaging (MRI)

Magnetic Resonance Imaging, also known as MRI, is an imaging procedure used for medical imaging purposes (Mayo Foundation for Medical Education and Research, 2019). MRI utilizes a magnetic field and radio waves, produced by a computer, to create precise images of the organs and tissues within the body (Mayo Foundation for Medical Education and Research, 2019). MRI machines are large, tube-like structures that contain magnets (Mayo Foundation for Medical Education and Research, 2019). A magnetic field is temporarily created inside the MRI machine which re-positions the water molecules in the body (Mayo Foundation for Medical Education and Research, 2019). Due to the presence of radio waves, the realignment of the water molecules are able to emit muted signals, which are used to produce cross-sectional MRI images (Mayo Foundation for Medical Education and Research, 2019). In addition, MRI machines also have the capability to reconstruct 3D images from various different angles (Mayo Foundation for Medical Education and Research, 2019). This non-invasive method of viewing the internal body is highly beneficial to doctors as it allows them to diagnose innumerable diseases, most often related to diseases of the brain and spinal cord, such as aneurysms, eye and inner ear disorders, multiple sclerosis, spinal cord injuries, stroke, tumours, and brain injury caused by trauma (Mayo Foundation for Medical Education and Research, 2019).

Discovery and Usage of MgB2 as a Superconductor

Superconductors also have a significant advantage in aircrafts (NASA, 2012). NASA's aviation goal's are to combine new aircraft configurations with an advanced turboelectric distributed propulsion system (NASA, 2012). This new creation is expected to increase the effective bypass ratio, while reducing drag (NASA, 2012). In order for this new entity to be successful, Jeff Trudell, an engineer at Glenn Research Center, says that cryogenic superconducting electric motors and generators are needed in order to lower fuel burn (NASA, 2012). Superconductors have the ability to do this because they can support much higher current densities, while also being small and light-weight in design compared to other room temperature materials (NASA, 2012). Superconductors also have the ability to conduct direct current without resistance, which means that less energy will be lost under a critical temperature and applied field (NASA, 2012). However losses due to alternating current comprise a significant portion of the heat load and is also based on the frequency of the current and applied field. To overcome these complications, a refrigeration system is needed in order to mitigate the losses (NASA, 2012).

A material known as magnesium diboride (MgB2) was found to be a superconducting material and ever since, scientists have been trying to discover methods to manufacture MgB2 at low-costs, while still maintaining the superconducting ability needed for their objective (NASA, 2012). The discovery of this material was quite impressive. According to Mike Tosmic, the president of Hyper Tech,"when MgB2 came along, we were excited because the transition temperature was twice as high as any known intermetallic superconductor at the time. We saw the potential for many applications" (NASA, 2012). Hyper Tech soon manufactured a set of four MgB2 rotor coil packs for a superconducting generator (NASA, 2012). This new material made of MgB2 was low in cost and density, and also had the ability to be easily configured into many different critical currents, while maintaining very low potential for alternating current losses (NASA, 2012). After this initial development, Hyper Tech continues to build a wide variety of MgB2 superconducting wires in various diameters and lengths (NASA, 2012).

Subsequently, MgB2 superconducting wire is now being used in MRI machines (NASA, 2012). By utilizing MgB2 superconducting wire in MRI machines for background coils, scientists hope to reduce the production costs of MRIs (NASA, 2012). Further, Mike Tomsic from Hyper Tech claims that, "[this is] the number one application for MgB2 wire" (NASA, 2012). Presently, MRI machines use niobium titanium superconductors that are cooled in a solution of liquid helium (NASA, 2012). The purpose of the liquid helium is to prevent the magnet from increasing in temperature due to overheating, ultimately preventing damage to the MRI machine (NASA, 2012). Unfortunately, liquid helium is a limited resource and the demand for liquid helium is expected to exceed the available supply (NASA, 2012). Due to this, the cost of helium has significantly increased over the past years, and it is predicted that the cost will continue to rise (NASA, 2012). As a result, this challenge has pushed the MRI industry to look for novel ways to convert to conduction cooled magnets (NASA,

2012). This method would allow the heat to be transferred from the superconductor coil to the copper links and then to a refrigerator, allowing for the removal of heat from the magnet (NASA, 2012).

The usage of MgB2 wires in MRI will influence the usage of MgB2 wires for other power applications such as superconducting fault current limiters, ultimately making other power applications more economical (NASA, 2012). Superconducting fault current limiters are useful when using renewable energy sources, such as wind power and solar power, because they prevent fault currents from exceeding the breaker capacity (NASA, 2012). Further, other innovations are in the works, "Hyper Tech and our collaborators have the potential to come up with a low cost fault current limiter. Full Size distribution voltage superconducting fault current limiters are scheduled to be on the grid in the next two years or less" (NASA, 2012). The potential applications of MgB2 superconducting wires is extraordinary as many countries, such as Asia, Europe, and the United States, have shown great interest towards them (NASA, 2012). Moreover, the increased utilization of MgB2 wires instead of existing traditional apparatuses will lead to "lighter, greener, low cost technologies" (NASA, 2012).

Conclusion

Superconducting magnets are very beneficial to society. The discovery of superconductors has ultimately been the reason for the development of high-level technologies, such as the Large Hadron Collider, NMR, and MRI, that exist today. Further, these technologies have allowed for a deeper insight into the field of science. Although the superconductors that exist today are astounding, scientists are still looking for ways to improve currently existing applications of superconductors, in hopes to uncover more ground-breaking discoveries.

References

Coalition for the Commercial Application of Superconductors. CCAS. (2014). http://www.ccas-web.org/superconductivity/nmr/.

Hockings, P. D., Hare, J. F., & Reid, D. G. (2010, March 17). Superconducting Magnets. Superconducting Magnets - an overview | ScienceDirect Topics. https://www.sciencedirect.com/topics/physics-and-astronomy/superconducting-magnets.

Kaseman, D., & Iyer, R. S. G. (2020, August 15). NMR: Introduction. Chemistry LibreTexts. https://chem.libretexts.org/Bookshelves/Physical_and_Theoretical_Chemistry_Textbook_Maps/Supplemental_Modules_(Physical_and_Theoretical_Chemistry)/Spectroscopy/Magnetic_Resonance_Spectroscopies/Nuclear_Magnetic_Resonance/Nuclear_Magnetic_Resonance_II.

Mayo Foundation for Medical Education and Research. (2019, August 3). MRI. Mayo Clinic. https://www.mayoclinic.org/tests-procedures/mri/about/pac-20384768.

NASA. (2012). Superconductors Enable Lower Cost MRI Systems. NASA. https://spinoff.nasa.gov/Spinoff2012/hm_6.html.

Pulling together: Superconducting electromagnets. CERN. (n.d.). https://home.cern/science/engineering/pulling-together-superconducting-electromagnets.

The magic of superconductors in the spotlight. CERN. (n.d.). https://home.cern/news/series/superconductors/magic-superconductors-spotlight.

Zinkel, B. (2020, June 19). Superconducting Magnet vs Permanent Magnets: What Are the Pros and Cons? Nanalysis. https://www.nanalysis.com/nmready-blog/2020/6/19/superconducting-magnet-vs-permanent-magnets-what-are-the-pros-and-cons.

6. Electric Applications of Superconductors

By Maria Gonzalez

Introduction

As previous chapters described, superconductors are an influential discovery in the scientific field that have many practical applications. This chapter will detail how superconductors can be used in power transmission. Superconductors are defined as materials that cease to have electrical resistance when cooled to exceedingly low temperatures (Encyclopedia Britannica, n.d.). The definition alone details why superconductors are so useful in the electric power industry—they have a lower level of resistance, and therefore suffer less energy losses than the regular conductors used in alternating current (AC) or direct current (DC) power transmission. A specific subtype of superconductors, called high temperature superconductors (HTS), are an especially attractive option for superconducting power transmission (SPT) because they widen the range of temperatures facilitating superconductivity. Not every material is capable of superconductivity, so a larger temperature range expands the list of possible superconductors and resultantly, what coolants can be used. With an increase in greenhouse gas emissions worldwide and consequently, the public's increased level of environmental awareness, the use of SPT globally is gaining traction because of its environmental benefits.

Low Temperature Superconductors (LTS)

Before the discovery of high temperature superconductors, low temperature superconductors were the norm in SPT. Superconductivity was discovered over a century ago now, with mercury being the first known superconductor (Marsh, 2009, p. 38). With a critical temperature of 4.2 kelvin (roughly -269 degrees celsius), mercury is considered a LTS, so its use for superconductivity is therefore incredibly expensive and highly complicated to operate (Marsh, 2009, p. 39). The most famous LTS system was called the Brookhaven system, and it ran from 1975 to 1986, at which point HTS were first identified (Jones, 2008, p. 4343). It consisted of a Nb_3Sn conductor cooled to its critical temperature of 18 kelvin by supercritical helium (Jones, 2008, p. 4343). The LTS cables never reached commercial success because they lacked practicality in terms of cost and complexity, and were quickly outdone by HTS after their emergence (Jones, 2008, p. 4343). Although LTS are typically an unfavourable option when compared to HTS, MgB_2 is an intriguing LTS (Jones, 2008, p. 4344). Firstly, it is a cheap superconductor, which is rare for LTS, and secondly, it is a lightweight material (Jones, 2008, p. 4344). Its critical temperature is 39 kelvin, so it requires temperatures in the range of 20-30 kelvin to function properly—this is feasible if a special refrigerator called a cryocooler is used, but this is an unrealistic option over long distances (Jones, 2008, p. 4344). Conversely, liquid hydrogen (which can reach 20 kelvin, proficient for the superconductivity of MgB_2) could be used for cooling purposes, but its use is forbidden by safety standards (Jones, 2008, p. 4344). Refrigeration does take a small amount of power which lowers the efficiency of the system, but in general, superconductor cables are highly efficient in comparison to other power transmission infrastructure (McCall, 2009, p. 56).

High Temperature Superconductors (HTS)

High temperature superconductors first appeared in 1986, opening the realm of possibilities for SPT—superconductivity could now take place at higher temperatures than ever before (Jones, 2008, p. 4343). Knowledge of HTS occurred after two scientists discovered that barium-doped lanthanum copper oxide reached superconductivity at 36 kelvin, around 12 kelvin higher than the next highest temperature superconductor known at the time (Marsh, 2009, p. 39). The discovery was influential—since the 1986 discovery of the first HTS, superconductivity at temperatures as high as 130 kelvin has been detected (Marsh, 2009, p. 39). For superconductivity at temperatures that high, liquid nitrogen is an accessible, sufficient coolant with a more reasonable cost than other cryocoolers (Marsh, 2009, p. 39). High temperature ceramic superconductors are typically made up of layers which include "two-dimensional copper-oxygen sheets along which superconduction takes place" (Encyclopedia Britannica, n.d.). The use of elevated temperatures removed some of the issues stemming from using LTS, including the expensiveness of certain types of LTS, and the limited number of superconducting materials in existence, which made them somewhat inaccessible (Jones, 2008, p. 4343). The discovery of

HTS outright eradicated the need for helium—an expensive substance in both liquid and supercritical forms—used for cooling superconductors to critical temperatures (Jones, 2008, p. 4343). Avoiding the use of helium also allowed for a wider temperature range during SPT, which is especially significant because even 20 kelvin could make a difference in reducing the potential for accidents (which were not unlikely when utilizing LTS) (Jones, 2008, p. 4343). Although other HTS exist, some notable examples are bismuth strontium calcium copper oxide (BSCCO) and yttrium barium copper oxide (YBCO) (Jones, 2008, p. 4343). Researchers and developers have faced difficulties while attempting to manufacture YBCO in long enough lengths for wires and tape, so they have shifted their focus to BSCCO (Marsh, 2009, p. 39). Fortunately, BSCCO has been one of the most reliable superconductors over the years of work on SPT, even capable of reaching lengths over a kilometre long, so it is a suitable option until YBCO wires and tape can be produced with more ease (Marsh, 2009, p. 39). It should be noted, however, that BSCCO's structure makes it frail and consequently inflexible; in order to construct cables, scientists have enclosed BSCCO in silver, which is a much more malleable material (Marsh, 2009, p. 39). In addition to making BSCCO more limber, a coolant is also required to achieve superconductivity, so a pipe is installed to allow nitrogen to cool the superconductor to its critical temperature (Marsh, 2009, p. 39). Beyond power transmission, superconductors can be incredibly diverse in their applications; they can be used inside generators, transformers, motors, or even power storage devices (Marsh, 2009, p. 39). Superconducting cables are also present in the famous Large Hadron Collider—the biggest particle accelerator on Earth (CERN, n.d.)—at CERN, the European Organization for Nuclear Research (Marsh, 2009, p. 39). The Large Hadron Collider uses liquid helium to cool the electromagnets formed by superconducting cables to a temperature of 1.85 kelvin, which is actually lower than that of outer space (CERN, n.d.).

Superconducting Power Transmission (SPT)

As mentioned in the introduction, people are becoming increasingly aware of their environmental footprints. Alongside concerns about the impact of their own actions on the environment, people are beginning to hold entire industries accountable for their careless destruction of the environment. The impending pressure of climate change and greenhouse gas emissions have decreased people's desire to rely on oil and gas and has led to an appeal for reform of the electric power industry. The prices of oil have also risen, making the use of oil and gas in the energy sector less desirable; furthermore, oil and gas are non-renewable resources that will inevitably run out (Jones, 2008, p. 4342). For these reasons and more, SPT is becoming a progressively more appealing option for energy production. Its benefits to the environment are impossible to ignore as greenhouse gas emissions continue to negatively impact the environment. The use of superconductors greatly improves the efficiency of power transmission—"3-5 times the power can be transmitted in a superconducting cable with the same, or less, loss than a conventional cable" (Jones, 2008, p. 4342). Because of the many advantages, the use of superconductors for power transmission has only grown. 2G, which stands for second generation, are conductors coated with YBCO, a type of HTS, forming cables that are cheaper to produce than copper cables (Jones, 2008, p. 4343). Different companies have been investing in the production of 2G wires—at the time of the article's writing in 2009, the newest version of 2G cables could transmit as much as 10 times the power of regular copper cables (Marsh, 2009, p. 39). There was also an instance in the United States circa 2002 where 8 tonnes of copper cable were changed out for 110 kg of superconducting cables—this resulted in a massive reduction in the amount of space taken up by power transmission infrastructure (Marsh, 2009, p. 39). HTS cables offer other improvements over regular copper and aluminum wires—they are capable of conducting roughly 150 times more electricity than ordinary wires of the same size, and this astounding power density is the reason behind their similar cost to overhead power lines (McCall, 2009, p. 55). SPT is effective enough to have gone global; companies in the United States, Holland, Denmark, and Korea are working on constructing underground power transmission networks between cities (Marsh, 2009, p. 39). Superconductors are also interesting in that they become fully resistive (thereby halting all conduction) if their "current carrying capacity reaches a natural limit, determined by magnetic and other factors rather than resistance" (Marsh, 2009, p. 39). This means that they are capable of containing fault currents, which is the case with the superconducting cables used by the company Consolidated Edison in their Manhattan power grid (Marsh, 2009, p. 39).

Although SPT sounds promising in theory, it also has its failings. One drawback of SPT is that it is better suited to DC power, and although DC power is superior for long-distance power transmission (a benefit for larger countries), AC power is what is utilized for actual power production and at the end for individual use (Jones, 2008, p. 4342). Additionally, attempting to convert the power type would be incredibly costly, so SPT in its current state is unlikely to be used exclusively (Jones,

2008, p. 4342). The power grid upgrades necessary for widespread SPT usage are discussed in the upcoming section.

The Environment and SPT

Wind Energy

As mentioned previously, the use of superconductors can greatly reduce the amount of energy lost to heat during power transmission. The efficiency of power transmission is boosted by the superconductor's removal of electrical resistance, because a specific quantity of energy is able to complete more work than it could while using a regular conductor (Marsh, 2009, p. 38). This means that with the exact same amount of energy, more things can be powered using a superconducting cable than a regular cable, making SPT an option better suited for protecting the environment from harm. This is because less greenhouse gas emissions need to take place for the transmission of a required amount of energy to occur (McCall, 2009, p. 56). As another "green" contribution, the installation of superconducting cables makes it cheaper for renewable energy to be used for power generation, which in turn further reduces greenhouse gas emissions (McCall, 2009, p. 57). Correspondingly, SPT can be applied to wind turbine farms, sources of renewable energy, which generate and transmit power over large distances to pre-existing power distribution networks (Marsh, 2009, p. 39). Developers at the American Superconductor Corporation (AMSC) are considering using superconducting cables— which they call "superconductor electricity pipelines"—to connect different inland renewable energy resources to largely populated cities on the coast (Marsh, 2009, p. 39). The use of superconductors in wind energy would likely considerably lower the cost of production because of the high power density (Marsh, 2009, pp. 40-41). Another advantage is that HTS can be used to construct offshore 10 MW wind turbines, which would weigh significantly less and take up much less space than regular 10 MW wind turbines (Marsh, 2009, pp. 40-41). The difference is telling—a reduction in weight from 300 tonnes to 120 tonnes, and only 3,000 to 4,000 10 MW wind turbines necessary to complete the work of 7,000 to 8,000 smaller wind turbines (Marsh, 2009, p. 41). The size reduction and increase in efficiency is especially important as the world's population and energy demands continue to grow. If SPT continues to succeed in the generation and transmission of wind energy, it could have the potential to make other renewable energy resources, such as hydropower, or current and wave power, much more effective (Marsh, 2009, p. 42).

Upgrades to Existing Power Grids

As the world's perspective shifts on humanity's detrimental effects on the environment, the governments of China, Japan, South Korea, and Europe are progressively investing more money into renewable energy (McCall, 2009, p. 54). However, in order for renewable energy to flourish, upgrades to existing power grids are required so governments can meet renewable energy goals (McCall, 2009, p. 54). To provide an example, a development project meant to erect wind turbines generating thousands of MW of energy in Texas was involuntarily suspended because the existing transmission infrastructure was inadequate (McCall, 2009, p. 54). Current power grids are outdated because they were built back when utility companies produced and distributed their own power, and any energy lost to heat during transmission was considered an acceptable loss because of the low cost of fuel (McCall, 2009, p. 55). As expected, the world's population and resulting energy needs have only increased since then, so the transmission of higher volumes of energy over longer distances has forced power grids to evolve accordingly (McCall, 2009, p. 55). Unfortunately, AC power transmission over larger distances leads to a greater amount of electrical resistance, and the quantity of AC power capable of being transmitted through an overhead power line decreases as the distance travelled increases (McCall, 2009, p. 55). Moreover, if AC power lines at a higher voltage than the existing system are installed, the majority of the existing power grid would have to be upgraded to bear the voltage difference (McCall, 2009, p. 55). Superconductor electricity pipelines, which were mentioned earlier, are a potential solution to the problem of power grids insufficient for extensive renewable energy transmission (McCall, 2009, p. 54). The superconductor electricity pipelines would be installed underground, would be only 3 feet wide, and would much more efficiently transmit power at a cost comparable to overhead power lines (McCall, 2009, p. 54). Superconductor electricity pipelines would transmit DC power and would be fitted with voltage source converters to allow transmission over large distances without much, if any, electrical resistance (McCall, 2009, p. 55). Some benefits of underground installation are that the cables are hidden—eliminating bulky overhead power lines disrupting the landscape—and they are also protected from any damage a powerful storm or deliberate attack could cause (McCall, 2009, p. 55). Once upgrades to existing power grades are finished, the electric power industry can finally devote itself to producing and distributing greener energy with less environmental damage.

Conclusion

In conclusion, superconductors are a beneficial addition to the electric power industry which have pushed SPT to the forefront of power transmission options. The increased efficiency from using superconductor cables—especially HTS—and the reduced impact on the environment makes SPT an attractive option for future power transmission infrastructure. SPT has been used in countries worldwide because of the greater amount of work it can complete with a certain amount of energy. SPT is not without its faults, however; some superconductors are expensive and restrictive, critical temperatures can be difficult to manage because of safety standards on cooling materials, and further research is needed to allow AC power to fully function with superconducting cables. Furthermore, existing power grids are unable to handle the transmission of renewable energy, which SPT is tied to. Once the issues with SPT are resolved, SPT may lead the electric power industry as the world becomes increasingly protective of the environment.

References

CERN. (n.d.). The large hadron collider. Retrieved June 24, 2021, from https://home.cern/science/accelerators/large-hadron-collider

Encyclopedia Britannica. (n.d.). Superconductors. Retrieved June 24, 2021, from https://www.britannica.com/technology/conductive-ceramics/Superconductors
Jones, H. (2008). Superconductors in the transmission of electricity and networks. Energy Policy, 36(12), 4342–4345. https://doi.org/10.1016/j.enpol.2008.09.063

Marsh, G. (2009). Rise of the superconductor. Renewable Energy Focus, 10(4), 38-42. https://doi.org/10.1016/S1755-0084(09)70152-8

McCall, J. (2009). Rise of the superconductor: part 2. Renewable Energy Focus, 10(5), 54-57. https://doi.org/10.1016/S1755-0084(09)70191-7

7. Key superconducting materials

By Darla Chloe Daniva

Introduction

Superconductors require materials that contain the properties to provide electrical conductivity without resistance at sufficiently low temperatures. It is important to recognize that all materials have different critical temperatures and will have to change properties differently when it comes to electron flow. One of these properties requires the material to have perfect diamagnetism (in comparison to semiconductors). This is also supported by what is called the Meissner effect, the famous phenomenon where any semblance of a magnetic field is expelled (Elprocus, 2021). Some other factors include not only the type of material but the length of the conductor, its voltage, and the surface area (Ndahi, 2003). These superconductors are classified into Type I and Type II materials. Superconductors hold materials that can be used in the engineering industry to create powerful electromagnets, computers, MRI (magnetic resonance imaging) and electrical systems. Different types of engineers explore the relationships between the critical temperatures and physico-chemical properties, such as resistance, in order to test for the most effective materials to be used in superconductivity. Materials must be at low energy states in order to turn superconductive, they must also be cold in order to be able to use a smaller amount of energy (which can be an environmentally-friendly advantage) (Elprocus, 2021). These classes of superconducting materials are always advancing. Superconducting materials can be, but are not limited to: alloys, ceramics, chemical elements, organic matter, and pnictides. Of course, there are scientists discovering new materials that can be classified as a superconductor. Unfortunately, not all materials will be suitable as a superconductor. For example, elements such as silver or gold can not reach the level of a superconductor as it must be able to withstand higher or lower critical temperatures (Ndahi, 2003). This chapter will maintain its focus on the current key materials and properties of the different classifications of superconductors, and where the materials can be used, along with several examples.

Type-I Materials

Type-I superconductors contain materials that can provide superconductivity at lower temperatures, in opposition to how type-II materials conduct. Examples of which material these types of superconductors can be are zinc, tin, mercury, or aluminium; these elements maintain a critical temperature that ranges from 0.88K to 4.15K (Ndahi, 2003). Besides these materials, niobium was found to have the highest critical temperature (Tolendiuly et al, 2021). Pure metals or elements are one of the common examples of these types of superconductors. They are also considered soft superconductors because of their ability to lose superconductivity simply in a critical magnetic field (Elprocus, 2021). NbTi and Nb$_3$Sn were considered to be a better generator for high magnetic fields than conventional resistant magnets, motivating further research on low-temperature superconductors (Ma & Yao,, 2021). In order to become classified as a type-I superconductor, the low critical temperature should be maintained around a range of 0K to 10K and obey the Meissner Effect (Electrical 4 U, 2020). The Meissner Effect is a phenomenon caused when a superconductor cools down to its critical temperature (approximately 0K to 10K), and causes a reaction where the magnetic field does not permit it to go through (Electrical 4 U, 2020; Elprocus, 2021).

The definition of superconductors and its properties means there should be multiple meta-analyses and experiments conducted. An experiment was conducted by Shopova and colleagues to study the Halperin-Lubensky Ma (HLM) effect of superconductivity on thin films, such as aluminium and tungsten (Shopova et al., 2003). The following experiment used the result of quasi two dimensional films and its strong HLM effect in three dimensional systems to support further research on the search for a key type-I superconductor film (Shopova et al., 2003). The experiment focused on the thermodynamic properties of the superconductors, revolving its attention on aluminium. This transition from a superconducting state to its normal state is described as sharp and abrupt (Electrical 4 U, 2020). The results of this experiment concluded that superconductors with thinner film, below 0.1μm (Shopova et al., 2003). However, the thickness of the aluminium film is purposely lowered for the HLM effect, and can be lowered to the point where superconductivity is destroyed in the specific experiment (Shopova et al., 2003). As a result, size of effect contributes to the maintenance of superconductivity (Shopove et al., 2003). These types of experiments are used to determine which materials are most efficient as a superconductor, as it is important to make sure all superconductors manufactured are not only problem-free but effective.

As mentioned above, one of the many factors that make an efficient superconductor is the length of the conducting material itself. To achieve a key length, Kozhevnikov and colleagues used an expression to find the Pippard coherence length of two extreme type-I superconductors. These superconductors made what is called a Cooper pair of electrons (Kozhevnikov, 2013). The results of this experiment demonstrated the ideal length of a superconductor, which varies depending on what material is being used.

For creating new materials while taking the expenses into consideration, NbTi continues to be the cheapest practical material because of the use of liquid nitrogen as a coolant (Ma & Yao, 2021). However, the cheaper solution is not always the most efficient. The NbTi superconductor has a more brittle material, however it is an issue which can be fixed with the process of wire fabrication (Ma & Yao, 2021). NbTi superconductors are actually the main material used for the manufacturing of the MRI and other systems, also making it one of the most consumed type-I superconductors (Ma & Yao, 2021).

Type-II Materials

High-temperature superconductors, also known as hard superconductors, are classified as Type II materials that have higher critical temperatures (Electrical 4 U, 2020). Unlike type-I superconductors, type-II superconductors only partly obey the Meissner Effect (Electrical 4 U, 2020). Type-II superconductors maintain a higher critical field (about 1T) and also exhibit two of these magnetic fields, rather than just one (Electrical 4 U, 2020). Examples of type-II superconductors are alloys and ceramics (Electrical 4 U, 2020). Other examples of type-II superconductors include NbTi and NBn. These key materials are effective for producing strong magnetic fields which can be used for the manufacturing of electromagnets (Electrical 4 U, 2020). Another unique characteristic these superconductors hold is that they can hold the existence of mixed-states, unlike soft superconductors (Ndahi, 2003). The properties of superconductor materials can be changed to achieve the desired characteristics of an efficient superconductor. Another key difference between type-I and type-II superconductors is the ability to withstand impurities and maintain superconductivity. Type-II superconductors have the ability to maintain superconductivity without being affected as significantly by slight impurities, which is a significant advantage (Ndahi, 2003).

Since hard superconductors must deal with the instabilities of critical temperatures, research must be conducted to examine relationships between temperature and the electromagnetic field in order to determine the best material for these superconductors. Taïlanov and Yakhshiev (2002) conducted a study to determine what causes the stability of type-II superconductors to become unbalanced, with the possibility of magnetic and thermal destruction in its interest. In its abstract, the experimenters claim the possibility that there is a direct influence of initial temperature and critical-state stability (Taïlanov & Yakhshiev, 2002). Knowing the scientific theories and methods of superconductors are highly important in making them efficient.

Hard superconductors have now typically been used in levitational applications, such as flywheel energy storage (Ndahi, 2003). Electrical and chemical engineers have also found this technology to be used in reducing wear-and-tear by the levitational properties of superconducting materials, saving the many expensive costs of maintenance and replacement of parts (Ndahi, 2003). Type-II superconductors also hold an advantage due to their high critical magnetic field, increasing its ability to be applied in technical applications (Ndahi, 2003). Aside from finding the best and studying the most common materials, there are more traditional materials and methods used in more complex applications. An example includes a superconducting motor, which uses copper coil to alternate the current (Chapman, 2000). This copper puck makes it possible for the motor to reduce energy loss, therefore reducing the energy consumption needed (Chapman, 2000). The shape of the material is typically cylindrical as it increases surface area, however, rectangular shapes have also been used (Chapman, 2000). For trains, hard superconductors can be used with a more linear version (Chapman, 2000). This was experimented on and developed by the University of Witwatersrand, Johannesburg. These copper rectangles are used to levitate the superconducting components of the train, with materials requiring current-conducting wire and iron armatures to create an oscillating current which would drive the train forward (Chapman, 2000). This method would then be named the maglev train project. These superconducting materials are usually static, but it is important to also have knowledge about the dynamic response they can cause. As stated earlier, the type-II superconductors produce two critical magnetic fields, one being the lower critical field and the other being the higher critical field (Chapman, 2000; Ndahi, 2003). These two magnetic fields contain the existence of the third state, which is a mixed state, as mentioned earlier in this chapter. The material used for supporting the mixed state, which is the partial penetration of the field into the superconducting material, is called the "flux tubes", which are thin filaments that are not superconductive (Chapman, 2020). These flux tubes are instead used to circulate around the non-superconductor, which then creates a vortex (also known as vortices), which gives the superconducting material existence to its mixed state (Chapman, 2020). This experiment concludes the reasons for the suggested hierarchy of models which can be used in discovering the behavior and creation of hard superconductors.

Although there is high complexity in developing material, especially for high-temperature superconductors, technical devices based on them are constantly being developed. The most popular hard superconductor is $YBa_2Cu_3O_7$ (yttrium barium copper oxide). As discussed earlier, superconducting materials are discovered by changing the necessary properties of materials to see if they are suitable and effective for use. This material can be manufactured in two different ways based on a solid-phase method (Tolendiuly et al., 2021). LaBaCuO (lanthanum barium copper oxide) and bismuth-based superconductors were also discovered as suitable material with the ability to carry larger supercurrents which are more often used for electro-technical applications as much cheaper substitutes (due to using liquid nitrogen as a coolant instead) (Ma & Yao, 2021). These opportunities create large-scale applications but also create challenges for making these superconductors. These bismuth-based superconductors are superconductors that apply to type-I or type-II materials. Another type-II superconductor is the copper-oxide material, which contains the property of maintaining a high critical temperature, a necessary property for high-temperature superconductors (Ma & Yao, 2021). MgB_2 which was discovered earlier in 2001, is also classified as a superconductor, capable of holding a high critical temperature as well, but is more promising for wire fabrication (holding two main methods for fabrication) (Ma & Yao, 2021). The disadvantage for this superconductor comes with weaker flux tubes, however, they do have a lower cost for raw materials (Ma & Yao, 2021). Superconducting pnictides include iron-based materials and interact with silver most efficiently, and at a lower cost (Ma & Yao, 2021). Fortunately, even though there are efficient and less expensive materials, consideration must be given to the environmental impact of creating and using these hard and soft superconductors.

Future Perspectives On Superconducting Materials

Superconducting materials are not only limited to low-temperature and high-temperature components. Fortunately, as the world's population dramatically increases, so will the need for hospitals and clinics. High-field implications like MRI and other large scientific projects are in high demand (Ma & Yao, 2021). What is even higher in demand, is the need to develop and advance these technologies, as they greatly challenge electromagnetic properties such as critical density or magnetic/thermal stability, which have been mentioned through experiments in this chapter (Ma & Yao, 2021). Of course, when these future ideas are implanted, technology maturity and costs will need to be considered in order to make the process of creating a superconductor with high performance easier (Ma & Yao, 2021). MRI and NMR systems are said to have grown in technology exponentially in terms of wire production (Ma & Yao, 2021). France is one of the countries exploring substitutions for the current superconductor, NbTi (Ma & Yao, 2021).

Superconductivity can also be applied to the power industry, with the use of maglev trains mentioned above in more populated regions. Additionally, when it comes to reducing the ecological footprint these power sources have, future research on conductivity is attempting to connect its power stations to adapt to the use of renewable energy (Ma & Yao, 2021). For example, other countries like the USA and Japan have contributed to the manufacturing of electric aircraft (E-aircraft). They have collaborated with NASA and Airbus to invest in a more eco-friendly source of transportation that would not only reduce greenhouse gas emissions, but also provide an aircraft built with materials that produce less noise (Ma & Yao, 2021). This is a significant advantage in the manufacturing of superconductors because power production is a common issue in pollution, especially in recent years. As of now, one of the top challenges appears to be the expensive costs of materials and manufacturing of superconductors (Ma & Yao, 2021). The future perspectives for superconductors hold many changes for all industries, including the science industry.

Conclusion

In choosing the key material for a superconductor, one must go over the several factors that determine its electron flow. The engineering and physics industries work together with others (such as the bioelectronic or medical industry) to provide the most efficient superconductors. New materials continue to be added to the list, depending on their properties, to classify as a type-I or type-II superconductor. Experiments continue to be conducted to find these key materials. These superconducting materials are put to work at places like hospitals, where MRI is regularly used. This necessity shows the importance in finding the most efficient materials in a superconductor. One flaw in superconductor technology can create an impact on issues like misinformation or causing dangerous reactions. The future holds opportunities for new materials to be used, as there are now room-temperature superconductors being discovered. These superconductors can use information gathered from the properties and manufacturing of type-I and type-II superconductors, and implant that knowledge into making newer,more efficient ones. In short, superconductors require key materials and choose the best manufacturing processes while also taking into consideration other factors, such as the environment, fuel efficiency, and expenses.

References

Chapman, S. J. (2000). A hierarchy of models for type-II superconductors. Siam Review, 42(4), 555-598. https://epubs.siam.org/doi/pdf/10.1137/S0036144599371913

Kozhevnikov, V., Suter, A., Fritzsche, H., Gladilin, V., Volodin, A., Moorkens, T., ... & Indekeu, J. O. (2013). Nonlocal effect and dimensions of Cooper pairs measured by low-energy muons and polarized neutrons in type-I superconductors. Physical Review B, 87(10), 104508. https://journals.aps.org/prb/abstract/10.1103/PhysRevB.87.104508

Ndahi, H. B. (2003). Manufacturing with superconductors. The Technology Teacher, 63(3), 17-21.https://go.gale.com/ps/i.o?id=GALE%7CA111304349&sid=googleScholar&v=2.1&it=r&linkaccess=abs&issn=07463537&p=AONE&sw=w

Shopova, D. V., Todorov, T. P., & Uzunov, D. I. (2003). Thermodynamic properties of the phase transition to superconducting state in thin films of type I superconductors. Modern Physics Letters B, 17(04), 141-146. https://www.worldscientific.com/doi/abs/10.1142/S0217984903005044

Tailanov, N. A., & Yakhshiev, U. T. (2002). On the stability of the critical state in hard superconductors with a heterogeneous temperature profile. Physics of the Solid State, 44(1), 16-21. https://link.springer.com/article/10.1134/1.1434476

Tolendiuly, S., Alipbayev, K., Fomenko, S., Sovet, A., & Zhauyt, A. (2021). Properties of high-temperature superconductors (HTS) and synthesis technology. Metalurgija, 60(1-2), 137-140. https://hrcak.srce.hr/246110

Yao, C., & Ma, Y. (2021). Superconducting materials: Challenges and opportunities for large-scale applications. Iscience, 102541. https://www.sciencedirect.com/science/article/pii/S2589004221005095

Electrical4U. (2020). Comparison of Type-I and Type- II Superconductors. https://www.electrical4u.com/comparison-of-type-i-and-type-ii-superconductors/

Elprocus. (2021). Superconductor: Types, Materials, and Properties. https://www.elprocus.com/what-is-superconductor-types-materials-properties/

8. Questions still being asked

By Aleefa Devji

Introduction

Superconductors, as we have learned, are capable of transmitting electricity without loss and allow for reduction in both the costs and greenhouse gases involved in power production as less power is required. At present, superconductivity is achieved through low temperatures, high pressures, and very low energy states of materials (Engineering 360, n.d.). As a result, superconductivity has only typically been possible with the utilization of very expensive, inefficient cooling processes. Research into superconductors is therefore focused on answering the questions of how to make superconductivity possible at higher temperatures, or even at room temperature. Throughout this section we will explore the limitations associated with superconductors, and how they are affecting the utilization of superconducting currents in the commercial industry in today's time. We will also begin to explore the questions researchers are asking and looking to answer as well as how they are experimenting for possible solutions to these limitations.

Why are Superconductors Not Used More Readily in Everyday Life?

Superconductors, although extremely efficient in electrical conductivity at low temperatures, are not used in everyday life or as readily as we would hope. The reason for this is that low resistivity is required for efficient conductivity, and most metallic electrical conductors that are used have finite resistivity - therefore, there are more losses due to heating (Ned, 2021). Superconductors were developed more than 100 years ago, and they were observed to exhibit no resistance to the flow of the electrical current that runs through them if the temperature at which it functions is sufficiently low enough which would make the resistivity exactly zero (Ned, 2021).

The problem therein comes from the use of superconductors for production of energy. Although the superconducting wire can allow current to flow indefinitely and induce a magnetic field that is stable when stationary, the issue occurs when the magnetic field is used for producing energy (Ned, 2021). When energy production is required, the supercurrent is affected, and the resistivity causes a reduction in the current. As a result of these issues with energy production and heat production that reduce the function of superconductive supercurrents, the application of superconductors is scarcely applied in everyday life (Ned, 2021).

How Does Superconductivity Work?

As discussed previously, superconductors were first discovered to conduct current at only extremely low temperatures and high pressures. This allows the electrons inside the material to cool down so much that they behave differently than they do at room temperature (Ned, 2021). This is what produces their superconducting state, a state that is completely new and behaves as no other. This is a phenomenon that is also not quite fully understood, how does superconductivity emerge in any of these superconductive materials? This is a question that is still being asked by researchers and the scientific community.

The only form of superconductivity that is currently understood is conventional superconductivity, although even this took 50 years for physicists to prove. In normal conductors, there is a crystal lattice made up of positive ions and the electrical current can be seen as electrons flowing through the crystal lattice (Ned, 2021). As these electrons flow, they are constantly colliding with ions and losing energy through heat production - the origin of resistivity in the typical metallic conductors that are not superconductive. The theory of conventional superconductivity is therefore explained by physicists to exist in some materials that can favour forming long-range pairs of electrons when functioning at low enough temperatures (Ned, 2021). These long-range pairs get tangled around the crystal lattice, and the web of pairs keeps the electrons from colliding with one another or with the lattice. Therefore, if there are no collisions in the material, there is zero resistivity to the flow of the electrons and no loss due to heat (Ned, 2021).

The theory that was used to explain conventional superconductors moreover predicted that this phenomenon could only exist at extremely low temperatures. As a result, superconductivity, although fascinating, would not be very useful in everyday life, because to keep the supercurrent flowing with zero resistivity, the system would have to constantly be cooled down to extremely low temperatures (Ned, 2021). Although this is the case, researchers and physicists are invested in understanding superconductivity and the possibility of achieving superconduction at higher temperatures.

Room-Temperature and High-Temperature Superconductivity

Until present time, one of the biggest questions still being asked about superconductivity surrounds the materials that are capable of productiving supercurrents and whether superconductivity is possible at higher temperatures. In the 1980s the first family of superconductors that were discovered to conduct supercurrents at higher temperatures were cuprates. This discovery led to the possibility that supercondivity may not be limited to only low temperatures as was thought previously (Ned, 2021). Cuprates, which are formed from layers of copper oxides, are currently the higher temperature superconductors known to exist under normal atmospheric pressure. The temperatures for superconductivity with cuprates can reach up to 133K (Ned, 2021). Some of the questions still being asked in superconductivity research surround what mechanisms in cuprates make this superconductivity possible because it is still not completely understood.

Research into superconductivity at room-temperature and above has been abundant throughout the years in the realm of physics. Until 2020 there were only a few records of achieving superconductivity effectively at high pressures and higher temperatures. In 2020 room-temperature superconductivity was successfully achieved for the first time at a pressure approximately two and a half million times the atmospheric pressure (Ned, 2021).

How Prevalent is Superconductivity in Nature?

Another area of research that surrounds superconductivity is its existence in nature. Although it is difficult to achieve for daily living and utilization for power and man-made processes, the existence of superconductivity in nature is not rare (Ned, 2021). When looking at the periodic table of elements, researchers have found that about 30% of the elements are superconducting under atmospheric pressures. When they include the elements that are only able to conduct supercurrents under higher pressure, 50% of the elements on the table can be superconductors.

The problems that physicists are finding in the realm of material research for superconductors is that the conditions for superconductivity to occur are extreme and different for different elements (Ned, 2021). They are also finding that some compounds are difficult to produce and are brittle or chemically unstable. This is an area of research that is still being explored to answer questions about which elements still exist that have not been found to conduct supercurrents and how they can be manipulated to be utilized under more favourable conditions.

Main Limitations to Superconductivity

Superconductors are used today as sources of strong and uniform magnetic fields because they produce much higher magnetic fields around the core of the electromagnet. The most advantageous and current use of these superconductors are MRI and NMR scanners which are helpful for individuals in the medical field (Ned, 2021). With that being said, there are limitations that exist in superconductivity and the commercial application for higher temperature superconductors have been limited as a result.

Questions are still being asked about molding of superconductive materials into mechanisms and man-made materials for superconductivity. For instance, into wires and technology that can be utilized in a commercial industry. Cuprates, the highest temperature superconductive material that we spoke of earlier, are difficult to manipulate and fabricate into materials such as wires due to their brittle properties (Ned, 2021). These complications result in much higher costs for production, but their ability to produce much higher magnetic fields and carry current makes them a focus for technological research.

Iron-based superconductors are another focal point of research as they are the second high-temperature superconducting family of materials (Ned, 2021). They were only discovered about 15 years ago, and as a result many questions are still being asked about their production, utilization, and application in the modern world of technological advancement. It is expected that because of their properties, they can be applied to bulk magnets, thin films, and wires (mechanism). Although this is the case, one limitation in the research and utilization of these iron-based superconductors is their possible chemical toxicity (Ned, 2021). Therefore, another question or focal point of research is how the chemical toxicity can be limited or if it is possible to utilize iron-based superconductors without posing other harms or risks for the environment and people.

As a result of the extreme pressures that are required to achieve superconductivity under higher temperatures, these superconductors are not viable for technological advancement and application yet. Many questions are still being asked about the future of superconducting materials and their future applications.

Current Questions Being Researched

The National Institute of Standards and Technology (NIST) has been actively researching various areas of superconductivity. As we know, there are many questions still being asked about superconductivity to understand the future applications that may be possible with the use of these materials and elements. The questions that are still being asked and researched encompass a broad range of materials related, pressure related, and temperature related conditions.

NIST research has been covering a wide range of interesting avenues for superconductivity related advancements in technology. With respect to the new high-temperature superconductors for instance, the research surrounds the determination of the physical properties such as elastic constants and electronic structure (Lundy, n.d.). They are also focusing on the development of new techniques such as magnetic-field modulated microwave-absorption and determination of phase diagrams and crystal structures of the elements and core magnet fields (Lundy, n.d.). In terms of the low-temperatures superconductor research, they are looking at the effects of stress on current density to fabricate new junction voltage standards (Lundy, n.d.).

Another area of research surrounding superconductors focuses on answering questions about field strength and the limitations surrounding the amount of current that a magnetic coil can carry (Preuss, 2010). This research has found that the amount of current carried by a magnetic coil is dependent on the physical properties of the superconducting material such as the critical temperatures and critical fields associated with them (Preuss, 2010). It is understood that to utilize superconductors and enable the accelerators of the future, they must develop magnets with a much greater field strength than is now possible, and so they must use different materials to achieve this (Preuss, 2010). There are research programs such as the LHC Accelerator Research Program or LARP that are working in conjunction with other laboratories to develop accelerator magnets built from other materials. One such material that they are working with is niobium tin (Nb3Sn), which is a brittle material that requires special fabrication processes to generate larger magnetic fields to achieve the goals that are set out for utilization of the magnets. Some of the materials that are most promising to produce these magnets are superconducting materials that work at high temperatures (Preuss, 2010).

As we have discussed, the utilization of high-temperature superconductors is difficult, and many limitations have been met with regards to their advancement (Preuss, 2010). In the realm of high-field magnet research though, the high-temperature superconductors are being experimented with to see if they can be used at low-temperatures. The experimental studies that are being done are showing that the superconducting materials such as Bi-2212 becomes superconducting at around 95K, and its ability to carry currents and generate high magnetic field increases as the temperature is lowered from that point down to about 4.2K which is the boiling point of liquid helium at atmospheric pressure (Preuss, 2010). These advancements in research are answering the questions of how superconductive materials can be manipulated with the use of more costly and easier to fabricate environmental factors. This would then be able to achieve the same effects that would otherwise require the utilization of extremely high-pressure environments (Preuss, 2010).

With the advancement of this research and the understanding of how critical points, temperatures and pressures can be manipulated to achieve more costly yet optimal conditions for superconductors further research can be conducted. With this research comes more questions about how to utilize, manufacture and implement these mechanisms for commercial use. The Very High Field Superconducting Magnet Collaboration has been working with several laboratories, universities, and industry patterns to develop new superconducting materials for high-field magnets (Preuss, 2010). Under the direction of these collaborations and associations, questions about the utilization of these still brittle materials such as Bi-2212. Although, one promising advancement is that Bi-2212 although as brittle as fine china, has been able to be manipulated into the form of round wires (Preuss, 2010). The advancement of research in these wires alone has begun to posit solutions for the limitations in research and commercial use of superconductors in everyday life. One of the biggest limitations was how to utilize the materials for use. These round wires are filled with particles of ground of Bi-2212 in a silver matrix, they are then heat treated to allow the particles inside to melt and form new textures upon cooling to weld the materials in the right orientation for conducting supercurrents. A huge advancement in the research of superconductors, although one that is still posing many new questions about the advancements and how they can be utilized commercially.

Conclusion

Superconductors are intricate, they are complicated, but they have so much potential for future advancement of technology and commercial systems. Their potential to produce power and reduce electrical power waste will be ground breaking, but there are many limitations to overcome and questions to answer to see the benefits from this technology. Superconducting materials are naturally occurring as we can see from the abundance of elements on the periodic table that are capable of superconducting currents in nature or under high pressure. The researchers and physicists involved in this research have spent decades trying to understand the mechanisms and properties that allow for the superconductive capabilities to be achieved. After much time, research, and observation some headway is being made, but the advancement of this technology is requiring an amalgamation of these experts and their individual understandings to make headway. With the collaboration of these laboratories and research facilities the research is paying off and questions are beginning to be answered. From the first efficient room-temperature superconductivity experiments in 2020, to the manipulation of the materials into wires, and the understanding of the material properties, research is heading in the right direction and it's looking promising. The possibilities for superconductive materials are endless, and it will be interesting to see how they will be utilized to efficiently produce electricity and reduce the electrical power loss associated with today's methods of electrical transfer.

References

Engineering 360. (n.d.). Superconductors and Superconducting Materials Information. Engineering 360. https://www.globalspec.com/learnmore/materials_chemicals_adhesives/electrical_optical_specialty_materials/superconductors_superconducting_materials

Lundy, D. R. (n.d.). A Brief Review of Recent Superconductivity Research at NIST. NCBI. https://www.ncbi.nlm.nih.gov/pmc/articles/PMC4943746/

Ned, A. (2021, February 25). Why Are Superconductors Still Not Used in Everyday Life? Predict. https://medium.com/predict/why-are-superconductors-still-not-used-in-everyday-life-409728093a5f

Preuss, P. (2010, September 10). Superconductors Face the Future. News Centre Berkeley Lab. https://newscenter.lbl.gov/2010/09/10/superconductors-future/

9. Public Perspective on Superconductors

By Lajendon Jeyakumar

Introduction

Looking through the periodic table there are many elements that carry different uses and purposes in parts of our daily lives. When it comes to power and energy through the conduction of electricity the importance of these elements are evident. Knowing that electricity is a vital component as part of developing civilizations and generations through advancements in technology and transportation, it is important to notice that there is a vital aspect involved in implementing electricity effectively into these tools and technology. As a result, the important aspect of superconductors is the missing piece needed to complete this "puzzle" of effectively using electricity into society. Electricity is known as a charge travelling from one end to another, however, in order to do so effectively it requires conductors to move the charges from one place to another. While conductors provide control over where the charges move, there are superconductors that aid with this process as well. Superconductors can be classified as either metallic alloys or elements that can produce electrical current to flow without energy loss under certain conditions such as being cooled to a specific threshold temperature (Jones, 2019). With the use of superconducting materials this also aids with the process of transporting electrons without any resistance or blockage to the desired destination. Along with the usage of superconductors, the process that allows for the flow of electrons without any resistance is known as superconductivity (*Superconductors and Superconducting Materials Selection Guide*, 2015). Superconductivity plays a major role in the actions that occur for superconductors, as this mechanism occurs when specific conducting materials are brought towards an object's critical temperature (*Superconductors and Superconducting Materials Selection Guide*, 2015). This occurs when the temperature decreases as well as the resistance decreasing, ultimately leading to the arrival of its critical temperature and disappearance of its electrical resistance (*Superconductors and Superconducting Materials Selection Guide,* 2015).

There are many instances where superconductor materials have occurred throughout our lives. Some common instances include lead, mercury, zinc, titanium, and tungsten as metals. Whereas, there are also nonmetal and metalloid examples such as boron, calcium, and silicon (*13 Examples Of Superconductor Materials,* 2018). With the use of superconductors, this prevents any resistance for electricity flow while also granting long range transmission of electrical currents to cover a larger geographic range in comparison to electrical transmissions done through regular conductors such as copper (Bianca, 2018). The issue of electrical efficiency has been a topic as more resistance with conductor materials often leads to less current flow. As a result, with the implementation of superconductive materials without limitational barriers in temperatures, this can eliminate the quantity of current wasted and be efficient in other applications such as transportation and medical equipment (Bianca, 2018). Knowing that electricity and efficiency are both important matters in not only Canada but on a global scale as well, examining the various perspectives within society in this chapter is important in not only gaining a better understanding and knowledge on the subject of superconductors, but also gaining awareness as to how society and the general public view the matter of superconductors and its effectiveness in modern day applications.

Significance and Relevance of Superconductors

Before tackling the ways in which superconductors most benefit society, it is best to compare how regular conductors show less effectiveness in comparison to superconductors. Conductors usually involve a piece of metal that helps move forward the electrical charge, an instance can be used with copper (a good conductor) leading to an electrical charge (Herbert, 2011). On the contrary, superconductors perform the same role as a conductor but it has no resistance to a charge resulting in no heat generated as well as no energy loss (Herbert, 2011). In an article done by Beth Herbert, she conducted interviews with William Halperin, who was an award recipient for the Initiative for Sustainability and Energy at Northwestern (ISEN), as well as John Evans, who was a Professor of Physics at Northwestern. What they both claimed through their knowledge of superconductors is its transmission range as well as electrical current efficiency (Herbert, 2011). In the article Beth Herbert stated " short distance transmission such as connecting to a computer can be effective regardless of normal conductors or superconductors, however in longer distance transmissions like from Niagara Falls into New York City this can result in a large loss of efficiency" (Herbert, 2011). This claim was further proven as 7% of electricity generated is lost in the process to New York City primarily due to the heat generated. As a result, many companies are looking into superconductors to help with wasted electrical current due to its benefits of less heat produced and low resistance. Superconductors can vary based on diverse conditions and as technological advancements occur along with more intriguing discoveries, there can be new innovations that can apply superconductor efficiency in various conditions.

The invention and discovery of superconductors, as well as their most effective materials, gave society a better understanding of implementing its use into more applications we utilize today in our daily lives. Superconductors are classified into two types: Type 1 and Type 2 superconductors. Type 1 superconductors are a kind of superconductor that implements basic parts for conductance for different facets such as microchips found in computers, and electrical cabling used to operate devices within homes (Superconductor:Types, Materials, Properties and Its Applications, 2020). These conductors are often less effective in comparison to Type 2 superconductors because they lack superconductivity when placed in the presence of magnetic fields (Superconductor:Types, Materials, Properties and Its Applications, 2020). Looking at Type 2 superconductors, these kinds of superconductors are able to handle high magnetic fields and still carry out superconductivity, but if it reaches a significant higher magnetic field than its device capacity this can result in a vortex state where the superconductivity will drastically decrease (Superconductor:Types, Materials, Properties and Its Applications, 2020). Superconductors are relevant in our everyday lives just without our complete awareness, they are famously applied in the field of medicine. Superconductors with the use of powerful magnets, allow for these medical machines to help scan the body's joints, brain, and spinal cord in order to find abnormalities within the body (Loeffler, 2019). It also has other future advancements pertaining to electricity efficiency as well as providing current technology to have more lifespans in terms of battery life as well as methods to make transportation more efficient as well (Loeffler, 2019).

With superconductors evolving with the current technology, it's important to understand both the benefits and its drawbacks. This can be really beneficial in solving public concerns surrounding electricity usage and ways for society to become more efficient. Producing electricity already comes with environmental drawbacks resulting from generation in power stations. As a result, by creating methods to conserve and use electricity efficiently, this can also correlate with more positive influences on the environment as well. Superconductors do have drawbacks that will be explained in greater detail in terms of the conditions it must fulfill, but with the progression of modern technology it is definitely possible to discover methods that can help remove the limitations for electrical current flow in only low critical temperatures instead of regular room or higher temperature values. Therefore, superconductors are significant for building electrical efficiency but also further building on existing technology to make them more effective.

Benefits and Controversy Associated with Superconductor Implementation

Superconductors have been incorporated in many essentials that have been utilized to provide humanity with the best possible living standards. The first benefit that superconductors have provided is in the development of medical technologies. One of the first applications for superconductors in the field of medicine was the utilization of magnets in magnetic resonance imaging (MRI). The use of MRIs is especially significant in viewing the internal condition of an individual's body and is useful in determining the next steps for many medical professionals. MRI gives doctors prior notice and allows them to see the inside of one's body without the need for surgeries (Blundell, 2011). In addition to providing the doctor the best knowledge of the patient's condition, the use of MRIs also help to detect for any tumours, neurological functioning of the individual, disorders in joints as well as muscles, and the status of heart as well as blood vessels (Blundell, 2011). The incorporation of superconductors into building MRI machines with the use of the magnets has not only advanced medical technology to help give detailed patient data about the individuals body, but it has also provides a detailed look into the individuals body to find any other issues that can be treated before later and critical stages (Blundell, 2011). As a result, although it is not a cure, the incorporation of strong superconductor materials and development of MRI technology has been crucial in saving lives from matters of life and death.

In addition to technology advancements especially in the field of medical sciences, another benefit associated with superconductor implementation is its efficiency in comparison to regular conductors. On a global-scale superconductivity is recognized as the "energy superhighway" as it provides more efficient distribution of electricity over long-distance transmission and its productivity can also be useful in economic trade as well (Hawsey & Morozumi, 2005). With the generation of electricity through methods such as fossil fuels, nuclear power plants, and hydroelectricity, this can contribute to environmental impacts such as air toxicity and this creates controversy as many countries face a lot of environmental drawbacks without the appropriate electricity gain (Hawsey & Morozumi, 2005). Essentially countries want an equal trade off with environment and electricity distribution so with the addition of both Type 2 and high-temperature superconductors this can be achieved. In more developed countries such as the USA, Japan, and Canada, electricity lost over long distance transmission accounts for over 8% due to resistance (Hawsey & Morozumi, 2005). So with this concern superconductors have been pondered upon as a possible alternative away from normal conductors. In order to overcome the drawback of superconductors relying on cooler critical temperatures, many countries have adopted hydrogen economies in order to aid the transmission processes and to prevent any electrical current loss in the process (Hawsey & Morozumi, 2005). In order to help provide better long-distance transmission of electrical current through cable wires using both gaseous and liquid hydrogen, liquid and gaseous hydrogen can help with removing obstacles such as heat and resistance due to its cooling properties (Hawsey & Morozumi, 2005). Incorporating this method not only provides the benefits of electrical efficiency and economic gain, but also can lead to a larger demand in handling electricity through methods of superconductors.

A big controversy that is associated with the capabilities of superconductors is surrounding its effectiveness in only cool temperatures and limited strength magnetic fields. In research conducted by University of Houston physicists, they believe the application of "Beans model" also known as "Critical State Model" does not show an accurate depiction of how superconductors really capture and hold onto magnetic fields (American Institute of Physics, 2016). In addition to this claim, Roy Weinstein, lead author of the study, and research professor at the University of Houston, states " superconductors and the uses of trapped magnetic fields helps with the creation of many modern technology such as electric lights, toasters, motors/ship engines, and medical scanners but with the misconception of magnetic fields this creates less advancements in current technology" (American Institute of Physics, 2016). What he means by this is that through his research, he already has found that as the magnet strength increases so does the performance of devices, this ultimately can mean that with more stronger magnetic fields this can also lead to reduction in materials to build the same product and save a lot more money (American Institute of Physics, 2016). Another intriguing controversy that was identified in the "Critical State Model" is how superconductivity is only seen as a slow and incremental increase, instead of sudden increases. Roy Weinstein believes that with future research to help get stronger magnets into superconductors this can help with efficiency of existing technologies compared to the "Critical State Model" which only relies on low critical temperatures and weaker magnetic fields (American Institute of Physics, 2016). Further research has also proven another controversy suggesting that it's not magnetic fields that are directly responsible for affecting the ability of superconductors, but the conductive materials that leak the streams of electrons from the current as temperatures change (Fishman, 2021). Therefore, there still exists controversy on the exact mechanisms of superconductors and what their exact properties are. In order to truly create ideal advancements with superconductors, it needs to be further studied by physicists so that it can be applied in a more efficient and useful manner without wasting materials and resources unnecessarily.

Societal Perspective on the Application of Superconductors

Society has been grateful for the benefits that superconductivity has provided in areas such as sustainable energy, low-emission transmission, and also aiding areas of medical research. Superconductors have a future in powering new and existing technologies in ways that can be efficient in terms of energy and economic standpoints. It has been suggested by research that global power usage will double over the span of the next four decades from 16 terawatt's (TW) to 30 TW (Nishijima et al., 2013). This creates greater societal importance for relying on superconductors in order to help and provide efficient superconducting power generation stations, stronger power transmission, and energy storage (Nishijima et al., 2013). The general population is beginning to grow faster with newer innovations, so it can be acknowledged that there will also be a significant increase in electricity generation due to the amount of dependency societies have on energy and electricity. Based on findings in research, with the production of electricity this can result in increased levels of greenhouse-gas emissions which have been a concern for quite some time (Von Wald et al., 2021). During the 2011 Equinox Summit being held in Waterloo Canada, it was examined that in order to stabilize the CO_2 levels in the atmosphere, energy creation needs to be met while also using non-carbon forms of energy (Nishijima et al., 2013). Therefore, it was suggested to actively use new superconducting energy systems to effectively store power from renewable sources (wind and solar power), while also implementing high-temperature superconducting systems seen in transformers, and heavy industry facilities (Nishijima et al., 2013). By doing so, it's expected that it can reduce up to 35-65% of greenhouse-gas emissions seen in the European Union according to the Kyoto Protocol (Nishijima et al., 2013).

Lastly, according to many researchers world-wide, another innovative way in which superconductors can help the environment is in matters concerning water pollution (e.g. litter in water). Based on findings, only 2.5% of the water on earth is fresh, and based on forecast models in the near future water bodies may become more dry as time progresses (Nishijima et al., 2013). A solution suggested by many leaders and scientists is using the magnets from superconductors to implement a groundwater treatment plan surrounding magnetic separation of contaminants so that way the water can be cleaned from harmful chemicals, and recycled as well as used again (Nishijima et al., 2013). Therefore, in terms of societal perspectives on the matters of environmental well-being, society believes that superconductors can aid countries in a more positive manner. With a great deal of environmental benefits, superconductors may be seen as a vital asset for solving issues pertaining to energy efficiency and environmental pollution.

Conclusively, another approach that society takes on superconductor implementation is seen in a both positive and negative manner. Superconductors are considered versatile as they are not only efficient from an energy standpoint, but they can be applied to make more productive versions of existing technology. With the use of the magnets, they are able to create magnetic resonance imaging (MRI) machines to detect body tissues and internal structures within the body to help with diagnosing patient injuries, they have also been applied in countries such as Japan and Germany for transportation to create levitating trains (Maglev) that have no resistance on the rails (The Electrochemical Society, 2015). With these many benefits this begs the question: what are the drawbacks? The drawback that occurs with superconductors is the demand for its materials and its costs for families and companies. According to an article by Flükiger (2012), when needing superconductors for large transmission or larger projects in general, it requires a larger quantity of conductive material. It was further mentioned that with superconductor material in the forms of cables, wires, and fault current limiters, they all have around 40-60 year lifetimes before they need to be replaced (Flükiger, 2012). This becomes an issue because although there are reduced costs of not doing underground projects by making larger and more powerful superconductive material and magnets, this can be expensive as it needs to be constantly replaced (Flükiger, 2012). As a result, there is an economic drawback down the road because it is uncertain to know how these costs can affect the overall economy since other countries may also be interested in the same materials for projects (Flükiger, 2012). Thus, although it is evident that superconductors can be life changing for society, due to a lot of countries all having the same idea this can create competition for materials and this market economy may have effects on pricing for families in the long run.

Conclusion

Society and many countries on a global scale have put a heavy emphasis on the matters of electricity and energy, and in order to solve these various issues superconductors seem to be the new innovation that helps countries move forward. Superconductors are a vital asset for helping with issues pertaining to energy efficiency as well as market trades with other countries. With many other benefits such as efficient transportation, as well as technological advancement in healthcare, superconductors may have more benefits that need to be further examined. Throughout this book, the objective is to give the reader a better understanding about superconductors, but also provide insights about its impact and effectiveness for future advancements. By the end of this text, hopefully the reader will get a greater sense about the ways energy is produced and how certain elements and conductive material influence the impacts of energy productivity and generation. Energy and electricity are both vital assets for humanity moving forward, and if superconductors have benefits that can outweigh the current economic and environmental struggles, then maybe it is the route that needs more attention and examination for generations to come.

References

13 Examples Of Superconductor Materials. (2018). LoreCentral. https://www.lorecentral.org/2018/09/13-examples-of-superconductor-materials.html

Andrew Zimmerman Jones. (2019). What Is a Superconductor? Definition and Uses. ThoughtCo. https://www.thoughtco.com/superconductor-2699012

Bianca. (2018, November 19). The Importance of Superconductive Materials. Despatch. https://blog.despatch.com/the-importance-of-superconductive-materials/

Blundell, S. (2011). Superconductors: What they're good for. New Scientist. https://www.newscientist.com/article/mg21228370-400-superconductors-what-theyre-good-for/

Fishman, Z. (2021, May 14). Controversial superconductor breakthrough disputed once again. The Academic Times. https://academictimes.com/controversial-superconductor-breakthrough-disputed-once-again/

Flükiger, R. (2012). Overview of Superconductivity and Challenges in Applications. Reviews of Accelerator Science and Technology, 05, 1–23. https://doi.org/10.1142/s1793626812300010

Global Superconductor Applications. (2015). The Electrochemical Society. https://www.electrochem.org/superconductors

Hawsey, R. A., & Morozumi, S. (2005). The Energy and Environmental Benefits of Superconducting Power Products. Mitigation and Adaptation Strategies for Global Change, 10(2), 279–306. https://doi.org/10.1007/s11027-005-9031-4

Herbert, B. (2011). The Superpowers of Superconductors | Helix Magazine. Helix.northwestern.edu. https://helix.northwestern.edu/article/superpowers-superconductors

Loeffler, J. (2019, February 15). Superconductivity has the potential to reshape the world, but what is it exactly? Interestingengineering.com; Interesting Engineering. https://interestingengineering.com/superconductivity-what-is-it-and-why-it-matters-to-our-future

Nishijima, S., Eckroad, S., Marian, A., Choi, K., Kim, W. S., Terai, M., Deng, Z., Zheng, J., Wang, J., Umemoto, K., Du, J., Febvre, P., Keenan, S., Mukhanov, O., Cooley, L. D., Foley, C. P., Hassenzahl, W. V., & Izumi, M. (2013). Superconductivity and the environment: a Roadmap. Superconductor Science and Technology, 26(11), 113001. https://doi.org/10.1088/0953-2048/26/11/113001

Physicists discover flaws in superconductor theory. (2016). Phys.org; American Institute of Physics. https://phys.org/news/2016-04-physicists-flaws-superconductor-theory.html

Superconductor : Types, Materials, Properties and Its Applications. (2020, February 5). ElProCus - Electronic Projects for Engineering Students. https://www.elprocus.com/what-is-superconductor-types-materials-properties/

Superconductors and Superconducting Materials Selection Guide | Engineering360. (2015). Globalspec.com. https://www.globalspec.com/learnmore/materials_chemicals_adhesives/electrical_optical_specialty_materials/superconductors_superconducting_materials

Von Wald, G., Cullenward, D., Mastrandrea, M. D., & Weyant, J. (2021). Accounting for the Greenhouse Gas Emission Intensity of Regional Electricity Transfers. Environmental Science & Technology, 55(10), 6571–6579. https://doi.org/10.1021/acs.est.0c08096

10. In popular culture

By Gabriela Ivanov

Introduction

In the previous chapters, we have learnt about what defines superconductors. As well as the history of the discovery of superconductors, the physics of superconductors, magnetic application, electric application, and many other key factors in understanding what superconductors are. In this chapter, we will focus on superconductors in pop culture. What may seem like a new phenomenon to some, superconductivity has been used in sci-fi movies, and it has also been discussed on platforms like TED Talks. Superconductors may seem like something that scientific research will have only spoken about, but they are also seen in pop culture. This chapter will discuss how mainstream media speaks about superconductors, and if our media is correct. This chapter will show how superconductors have been represented in movies like Avatar and the Terminator franchise, in TV shows, as well as even in video games. It is important to note that in our current day, superconductors are not extremely mainstream in pop-culture, but they have been mentioned in some of the most mainstream movies of all time. In the upcoming section, we will discuss the use of superconductors mentioned in TED Talks.

Ted Talks

A TED Talk is a video created from a presentation at the main TED (technology, entertainment, design) conference or one of its many satellite events around the world (Wigmore, 2014). TED talks have become famous for giving a platform to some of our world's greatest intellects. Topics covered in TED Talks can range from activism to science, and hundreds of other topics in between. In 2011, Boaz Almog hosted a TED Talk on superconductivity. Boaz Almog is a physicist, scientist, and CEO of Quantum Experiences Ltd. Almog demonstrated how a superconducting disk can be trapped in a surrounding magnetic field to levitate above it, which is a phenomenon known as "quantum levitation" (Almog, 2012). Prior to conducting his experiment, Almog defined superconductivity as, "a phenomenon of exactly zero electrical resistance and expulsion of magnetic fields occurring in certain materials when cooled below a characteristic critical temperature" (Almog, 2012). He then continued to explain how even though superconductors and superconductivity were discovered over 100 years ago, only due to technological advancements, was he able to perform this experiment (Boaz Almog, 2012). Almog then performed his experiment to the audience. In this experiment, he placed an object 70,000 times heavier than a super thin 3-inch disk, in which the disk began to levitate said object. This was done through superconductivity, through which he describes is only possible with zero electrical resistance, and the expulsion of magnetic field from the interior of the superconductor (Almog, 2012). In physics, electricity is the flow of electrons inside a material and whilst these electrons are flowing they collide with atoms, and in these collisions they lose a certain amount of energy and then produce heat (Almog, 2012). However, inside a superconductor, there are no collisions, so there is no energy dissipation (Almog, 2012). Almog believes this phenomenon is quite remarkable because in classical physics there is almost always some sort of friction (Almog, 2012). His experiment was possible because superconductors do not "like" magnetic fields. So, a superconductor will try to expel magnetic fields from the inside, and it has the means to do that by circulating currents (Almog, 2012). Now, with the combination of both effects - the expulsion of magnetic fields and zero electrical resistance - creates a superconductor (Almog, 2012). Although as mentioned before, superconductors do not "like" magnetic fields, this experiment was made possible because a miniscule amount of magnetic fields remained in the superconductor (Almog, 2012). When the superconductor was placed on top of a magnet and began to levitate it remained still, this was caused by the discrete amount of magnets that remained inside of the superconductor. If we think about two magnets that repel each other, there is usually some sort of movement that occurs. These magnets will shake back and forth, but this superconductor remains completely still whilst levitating. Almog even moved the magnet to the left and right, and flipped it upside down and the superconductor remained still. This stillness can be defined as quantum locking, quantum locking occurs when a superconductor becomes trapped in a magnetic field, it will then be trapped in space and will not move without outside force (Quantum Lock, 2020). Almog then placed the superconductor above a circular magnet, and the superconductor began to rotate in a circular motion, this was because the magnetic produced an even circular magnetic field for the superconductor to rotate and levitate without friction (Almog, 2012). Almog then showed the superconductor levitating on a bigger platform. This bigger platform was a circular magnetic platform much larger than the previous circular magnet. The superconductor then began to levitate freely around this platform while also resembling a spaceship as it glided effortlessly without friction. Even after Almog flipped the platform upside down the superconductor continued to levitate and spin around the large magnetic platform. Almog then begins to explain how superconductors are used in our everyday lives. Through the use of superconductors we have been able to create MRI machines (which will be defined in the following section), as well as power lines, and power stations that can be powered by one superconducting cable (Almog, 2012). Furthermore, if this small disk that was half a mm thick could maintain the weight of something 70,000 times its own weight, a disk that is 2mm thick could carry 1000 tons which is approximately equivalent to the weight of a small car. Future directions of superconductors will be further explained in chapter twelve. It was through this experience that we have been superconductors in pop culture in one way; through TED Talks. To view this TED Talk and its thorough demonstration of this amazing experiment, simply enter Boaz Almog 'The Levitating Superconductor' into a search engine .TED talks are educationally informative, but they also still lie in the realm of social media culture. Superconductors have also been used in movies and TV shows.

In Film

As well as in TED talks, superconductors have been used in some mainstream films. It was even used in one of the highest grossing movies of all time; Avatar. Avatar is a Sci-Fi movie released in 2009 that is based on the fictional planet Pandora (Avatar, n.d). In summary,

> "When his brother is killed in a robbery, parapalegic Marine Jake Sully decides to take his place in a mission on the distant world of Pandora, there he learns of greedy corporate figurehead Parker Selfridge's intentions of driving off the native humanoid "Na'vi" in order to minde for the precious material scattered throughout their rich woodland. In exchange for spinal surgery that will fix his legs, Jake gathers knowledge of the Indigenous race and their culture, for the cooperating military unit spearheaded by gung-ho Colonel Quaritch, while simultaneously attempting to infiltrate the Na'vi people with the use of an "avatar" identity. While Jake begins to bond with the native

tribe and quickly falls in love with the beautiful alien Neytiri, the restless Colonel moves forward with his ruthless extermination tactics, forcing the soldier to take a stand- and fight back in an epic battle for the fate of pandora" (Avatar, n.d).

Now that there is an adequate summary of the film provided to investigate, we will now discuss how superconductors were implemented through the film. On the planet of Pandora lies a mineral named Unobtanium. Unobtanium is a room temperature superconductor for energy, which makes it very valuable; it is worth $20 million per kilogram unrefined and $40 million per kilogram refined on earth (Unobtanium, n.d). Unobtanium caused a great amount of grief between humans and Na'vi relations because humans mined Unobtanium for energy generation on Pandora, but the Resources Development Administration (RDA), began to suppress this development (Unobtanium, n.d). The atmosphere in Pandora was very toxic for humans which made it very expensive to mine and return back to earth (Unobtanium, n.d). Humans would transport Unobtanium on specific trucks the movie mentions as Hell Trucks from the mining sites back to Hell's Gate for refining transport to Earth (Unobtanium, n.d). Due to the need of dramatizing superconductors for cinematic effect, the superconductors in the film actually had an extremely strong magnetic field, which reversed prior knowledge (which we know to be true), that all superconductors repel magnetic fields (Unobtanium, n.d). The unusual property of the Unobtanium superconductor that it has a strong magnetic field, was set in place to levitate huge islands (Unobtanium, n.d). These huge islands were named the Hallelujah Mountains by "Earth's explorers", but on the contrary were called Thundering Rocks by the Na'vi, who hold them sacred (Unobtanium, n.d). The magnetic properties are also used to propel ships like ISV Venture Star, which is one of the interstellar ships used to transport supplies between earth and Pandora (Unobtanium, n.d). Without Unobtanium, interstellar commerce on such a scale would not be possible (Unobtanium, n.d). Unobtanium was not only the key to Earth's energy needs in the 22nd century, but it was the driving force of interstellar travel, and the mineral that kept the Hallelujah mountains afloat (Unobtanium, n.d). Although Unobtanium is rather expensive to mine, the more that is mined the more ships can be built and the more equipment can be sent to Pandora for further mining (Unobtanium, n.d). It is clear that even in fictional settings like Avatar, the movie and its story plot conveys the message that the use of superconductors are extremely important.

Avatar was one of the most famous films to use superconductors as a driving force of the film, but it was also even mentioned in The Terminator. A basic synopsis of the film is that the Terminator is a 'cybernetic organism', robotic assassin and soldier, designed by the military supercomputer Skynet for infiltration and combat duty, towards the ultimate goal of destroying the human race (T-800, n.d). In the fictional world of the Terminator franchise, The Terminator's Central Processing Unit (CPU) was a room-temperature superconducting artificial neural network with the ability to learn. Essentially, the Terminator was a machine that could kill in an instant, but it wanted to be more friendly and that required a great deal of power and energy, hence the use of superconductors. Although, the room-temperature superconductor actually caused a nuclear holocaust after it was used to create the first sentient computer, Skynet (Unobtanium, n.d). Although tragic, this effect did sufficiently represent the great powers that superconductors have.

It is interesting to note that superconductors are not extremely mainstream, but they were very prevalent in two of the most well known movies of all time. Avatar grossed over 2.8 billion dollars in the box office, and the Terminator franchise is infamous. Both of these movies were essentially based around the scientific methods of superconductors. In Avatar, Pandora and Earth had feuds about Unobtanium, as well as Pandora's whole island essentially being a superconductor. In the Terminator 2, superconductors were used to create human-like emotion in the Terminator. The use of superconductors in both movies may be because both movies had the same director; James Cameron. Mark Sappenfield even believed that Cameron, "does not write science fiction, he writes science fact" (Sappenfield, n.d). This is because Cameron's use of science in his films, as well as superconductivity, may be exaggerated for cinematic appeal; but the science holds true. In the next sections, we will discuss superconductors in television, as well as its representation in video games.

Superconductors in TV

Recently, there have been an abundance of hospital related TV shows. One of the most popular series by far is Grey's Anatomy. In this show there are many scenes involving Magnetic Resonance Imaging or otherwise known as; MRI machines. MRI machines actually use niobium titanium superconductors that are cooled in a bath of liquid helium, the liquid helium helps prevent magnet quenches where the magnet increases in temperature due to local overheating and can cause damage (Superconductors Enable Lower Cost MRI Systems, 2012). An MRI is a non-invasive imaging technology that produces three dimensional detailed anatomical images, it is often used for disease detection, diagnosis, and treatment monitoring (Magnetic Resonance Imaging, n.d). So, any scene in a hospital related TV show in which you see an MRI machine, there are superconductors at work.

Other than MRI machines, superconductors have rarely been mentioned in television. This may be due to the fact that superconductors may be a challenging concept for some, and it is hard to explain in a 30 minute television episode. Hopefully in the future there will be more mention of superconductors, or possibly these hospital TV shows may one day explain how MRI machines work, and then superconductors can be mentioned.

In Video Games

In 1997, there was a video game created entitled Wipeout, in which the player pilots a futuristic anti-gravity racing ship around utopian race circuits and the game's timeline was set in 2052 (Ingram, 2012). In 2012, a similar game was created but it was instead based off of the premise of superconductivity and quantum levitation (if you remember, this is what the TED Talk was mainly focused on). The mini-wipeout game works using, "quantum levitation- when supercooled (below 301 F), an electrically and magnetically neutral object coated with a ceramic layer - in this case, the ships - becomes a superconductor. This is an object that conducts electricity with no resistance, and no energy loss (Ingram, 2012). This game shows how superconductive objects and magnetic fields repel, which is caused by the Meissner effect (Ingram, 2012). Essentially, the ships in this game move extremely precisely due to them being superconductors. If you remember back to Almog's experiment, it is exactly the same as the way the sphere levitated and glided around the large magnetic structure, but this is instead in video game form. In this video game, the use of superconductors is extremely useful, as it differs from other race car games where the cars shake, and it is hard to steer your device. Superconductors would be very useful in the video game world, as video games are meant to be a form of escapism, and superconductors create a fantasy because they seem as if they are very technologically advanced.

Conclusion

In this chapter we spoke about how superconductors are seen/used in popular culture. In section one, we learnt about Boaz Almog's experiment with superconductors and quantum levitation. This section showed how his experiment was created, as well as provided further knowledge on superconductors as a whole. In section two, we learnt about superconductors in movies. Specifically, in Avatar and the Terminator franchise. In Avatar, the whole city of Pandora is fueled by the superconductor Unobtanium, and humans on earth are trying their best to get their hands on some. In the Terminator, the Terminator's CPU is run by a room-temperature superconductor, to try and give him some human emotion. In television, we can now know that superconductors are used in MRI machines, and in any television show where MRI machines are shown, superconductors are at work. In video games, we see a game that uses superconductors to create a smooth ride for ships and cars. Superconductors are not as mainstream of a topic as one would think, with minimal coverage in pop culture. Hopefully, one day everyone will know what superconductors are, and how interesting of a concept they are.

References

Almog, B. (2012). The levitating superconductor. TEDtalks.com. https://www.ted.com/talks/boaz_almog_the_levitating_superconductor/transcript#t-588080

Hewage, R. (2020). "Quantum Lock" and the Future of Application Security. HackerNoon.com.https://hackernoon.com/quantum-lock-and-next-super-secure-applicationsquantum-lock-fd1w321e

Ingram, A. (n.d). Quantum Levitation Racing: Wipeout VideoGame Made Real. Motor Authority. https://www.motorauthority.com/news/1071273_quantum-levitation-racing-wipeout-videogame-made-real

Sappenfield, M. (2009). Avatar: the real-life science behind the fantasy. The Christian Science Monitor. https://www.csmonitor.com/Science/2009/1228/Avatar-the-real-life-science-behind-the-fantasy.

Wigmore, I. (2014). Ted Talk. Whatisit.com. https://whatis.techtarget.com/definition/TED-talk#:~:text=A%20TED%20talk%20is%20a,may%20be%20on%20any%20topic.

Avatar. (2009). IMDB. https://www.imdb.com/title/tt0499549/plotsummary

Magnetic Resonance Imaging (MRI). (n.d). National Institute of Biomedical Imaging and BioEngineering. https://www.nibib.nih.gov/science-education/science-topics/magnetic-resonance-imaging-mri

Superconductors Enable Lower Cost MRI Systems. (2012). Nasa Technology Transfer Program. https://spinoff.nasa.gov/Spinoff2012/hm_6.html#:~:text=Tomsic%20explains%20that%20MRIs%20currently,overheating%20and%20can%20cause%20damage.

T-800. (n.d). Fandom. https://neoencyclopedia.fandom.com/wiki/T-800

Unobtanium. (n.d). Fandom. https://james-camerons-avatar.fandom.com/wiki/Unobtanium#:~:text=5%20References-,Description,it%20is%20expensive%20to%20mine.

11. Future directions

By Sifar Halani

Introduction

The energy demands of modern life in developed countries has increased significantly over the years as technology has come to play an increasingly large role. From lighting and heating to transportation and computing, the very foundations of our modern civilization are embedded in the ongoing availability of, and access to, energy. Global energy consumption has increased from just over 5,600 terawatt-hours in 1800 to over 171,000 terawatt-hours in 2019 (Ritchie, 2020). Both China and the USA alone each use more energy per year today than the entire world did during World War II. As this energy consumption rate continues to rise exponentially, we will need more efficient methods of transferring and using energy in order to maximize the value gained by the generation of energy.

Efficient energy use has the potential to make a greater impact on our world's energy reserves and humanity's environmental impact than the oncoming transition to renewable energy. Currently, around 65% of all available electric energy is lost in the conversion process from raw materials to electricity on the grid (Wirfs-Brock, 2015). Following this, an additional 6% is lost when transporting the energy from the powerplant to one's home. This means that in reality, only a small portion of the total available energy is actually being delivered to a household for consumption. Due to this energy loss and inefficient conversion and transportation, more energy must be produced in order to make up for the amount lost. If we were able to convert and transport energy without loss and use the full 100% of energy available, we would have to produce much less electricity and use far fewer raw materials overall. This means less greenhouse gas emissions and fossil fuel use. The positive impacts from full efficiency could outweigh the positive impacts from the slow, ongoing transition to renewable energy sources.

Superconductors have the potential to make full efficiency possible. This technological shift will have massive impacts on electricity supply and transportation, among other fields. As scientists discover newer and more efficient superconducting compounds, the number of feasible applications for superconductivity will continue to rise. This chapter will discuss how superconductors will change electricity supply and transportation, and also discuss a few other interesting future applications of superconductivity.

Electricity Supply

Power lines are widespread throughout the modern landscape. These huge, skeletal, metal structures that carry electricity from the power plants to our communities occupy vast swaths of land and create hazards for the surrounding areas (New Health Advisor, 2020). Power lines have been claimed to cause leukemia, cancer, and depression, and proven to cause severe harm or even death through electric shock. Power lines do not make up for these risks with any strong benefits. They still cause power to waste away during transport due to heat and light and affect property values negatively as well (New Health Advisor, 2020). Superconducting transmission lines are an upcoming concept that could replace traditional power lines in the near future. By using superconducting materials, these transmission lines overcome many of the downfalls of traditional power lines (Thomas, et al., 2015).

Superconducting transmission lines have a much higher capacity for energy transfer compared to traditional power lines, and do not suffer from the same losses due to their inherently efficient nature. The only losses generated by superconducting transmission lines are those associated with cooling them to maintain their operational temperatures (Thomas, et al., 2015). An option for reducing this loss even further is to place superconducting transmission lines underground as opposed to above ground. Being underground naturally provides a cooler environment for the lines, reducing the energy cost of keeping them cold. Being underground also has numerous additional added benefits. For one, there is no visual impact on the landscape. There will be no need for large, metal towers occupying space because the transmission lines will be underground. This frees up the option to replace existing above-ground power transmission infrastructure with alternatives such as farmland, housing, or commercial buildings (Thomas, et al., 2015). As our cities continue to expand and grow, this extra available space will be invaluable for social and economic development. The obsolescence of power towers also means less of an overall environmental impact, as tunneling does not require as many materials, and superconducting transmission lines can also be run alongside or within existing tunnel systems as well. This is possible because a 10-centimetre-wide cable of superconducting material can transmit as much electricity as a 17-metre-wide trench of standard power cables (Thomas, et al., 2015). Superconducting lines also do not raise the heat level of soil, causing less of an impact on ground fertility and humidity.

Another positive aspect of superconducting transmission lines is that they reduce property acquisition, saving costs for the government and causing less disruption for civilians and homeowners. The local real estate market remains unaltered because the superconducting lines are not visible above ground (Thomas, et al., 2015). If anything, the benefits of these new power lines would increase the value of nearby properties. Traditional power lines can often be damaged by natural phenomena such as rain, wind, snow, and ice. Superconducting lines would not share this susceptibility to damage due to their placement underground, making electricity supply more reliable for people, even during extreme weather events (Thomas, et al., 2015). Minimizing downtime on the power grid is especially important in regard to greenhouse gas emissions. When the power goes out in one area, the production loss must be compensated with increased power from other generators and grids, leading to increased emissions (Thomas, et al., 2015). This creates a drain on regional and even national power networks. In some cases, the annual energy loss from power line downtime adds up to the equivalent of multiple additional coal-fired power plants operating. Traditional power lines create a low humming noise and have been known to emit electromagnetic radiation fields that can cause negative health effects for those living nearby (New Health Advisor, 2020). Superconducting lines are silent, because they are completely efficient, and emit fewer radioactive waves than traditional lines (Thomas, et al., 2015).

Renewable energy also stands to benefit from the implementation of superconductivity. Energy sources such as solar and wind power are reliant upon nature to determine the schedule and rate at which they generate electricity. As such, there are often times when the power these sources generate goes unused due to low demand (Thomas, et al., 2015). An example of this would be a windstorm in the middle of the night where a large amount of electricity is generated by wind turbines but is not used because most people are asleep. This energy would then have to be stored in large batteries or sent off to other areas where it would be needed. Using current storage and transfer technologies, much of this excess electricity would end up being lost (Thomas, et al., 2015). This lack of control over when power is generated and inevitably wasted detracts governments and corporations from adopting renewable technologies. Using superconductors, excess energy generated during off-peak usage times could be stored in an endless loop without any wastage and then accessed when required. Power could also be transferred wherever else it may be required without any loss, as well (Thomas, et al., 2015).

Wind power is particularly suitable for upgrades using superconductivity. Wind turbines are currently large structures that occupy huge areas of land known as "wind farms". They require these large spaces in order to take full advantage of the wind blowing through (Islam, et al., 2014). Research is currently underway on superconductor-powered wind turbines that may be rolled out in the coming years. These new wind turbines will use superconductor-based generators that will be much smaller and weigh far less than traditional copper-conductor generators (Islam, et al., 2014). This will lead to the development of compact, lightweight wind turbines that will occupy less space on the land, require fewer materials to construct and maintain, and operate far more efficiently.

Electricity is not only a major part of the foundation of modern life, but also a benchmark for comparing quality of life across nations. There is a direct correlation between electricity production/consumption and standard of living (Grant, 1997). Countries that use high amounts of electricity tend to have higher standards of living and greater GDP per capita. As nations across the world approach higher standards of living, the energy being demanded across the world is increasing considerably (Grant, 1997). This relationship can also be flipped, as the availability of ever-greater quantities of power could also be a force allowing nations to obtain higher standards of life. Current solutions to this challenge involve massive networks of gas and oil pipelines that transport raw materials for conversion to electricity in power plants across the continents. Another approach has been to use the gas and oil to generate electricity on-site and then transport it using long chains of power cables rather than pipelines. Both of these methods incur huge losses and environmental degradation, in addition to land use (Grant, 1997). An alternative to this is a global network of superconducting cables known as a "smart grid".

In the event that our fossil fuel supply begins to dwindle, or global warming makes fossil fuel use no longer viable, researchers say that we may convert to a more centralized power structure where electricity comes from fewer, but larger generator sites (Grant, 1997). This would require a different type of grid to accommodate the changes in the generation system. A smart grid would be able to take advantage of the various factors associated with renewable energy sources and combine them with other sources, such as nuclear power, in order to create stable energy production and delivery across its covered area (Mikheenko & Johansen, 2014). It would accomplish this task using superconducting generators and power cables, making it a "supergrid" as well as a smart grid. Supergrids allow for efficient long-distance energy transfer, making a global smart grid a possibility. Such a grid would allow renewable energy sources to be used to their maximum potential. Solar panels could be placed in heavily sunlit areas such as deserts, wind turbines could be placed in the windiest valleys and seas, and tidal generators could be placed alongside beaches anywhere in the world. The power generated from these ideally placed renewable sources would then be organized and stored on a global system which would then distribute this energy wherever in the world it would be needed (Mikheenko & Johansen, 2014). The distance to and from the actual energy generation sites would be irrelevant because no energy would be lost during transportation anyway. Unused energy would be stored efficiently on a superconducting loop until it is needed. While the technological feasibility of a global smart supergrid could be attained in the near future, the socio-political and economic barriers to such a large-scale endeavour could prove more difficult to overcome.

Transportation and Other Applications

A promising application of superconductivity is in magnetically levitated transportation. Currently, there are numerous Maglev trains in operation across the world, with many more under construction (Whyte, 2016). The first such train launched in Shanghai in 2004. In the years to come, it is expected that Maglev trains will steadily replace other existing rail transport systems. In a Maglev train system, superconducting magnets push the train off the ground while other magnets placed around the rail and tube push the train down, left, and right, resulting in the train hovering in a central location (Whyte, 2016). There are then other magnets aligned along the walls which pull the train forward. Because the magnetism is constant and there is no loss with superconducting materials, these trains require minimal energy to operate and could be run with no fossil fuel consumption or greenhouse gas emission. They have the added benefit of extreme speeds up to 375 miles per hour, while still causing less turbulence than traditional trains. (Whyte, 2016). Another aspect of

Maglev trains is strong safety. There is no driver in a Maglev system since the magnetic walls will guide the train wherever it needs to go. Because the walls maintain a constant magnetism in one direction, thereby creating a constant speed for the trains running on them, one train cannot intersect paths with, or run into another one. Derailment is also highly unlikely due to the nature of the magnets. The further a train gets from its central hovering position, the stronger the pushing force from the respective magnets will be to get it back to its centre position (Whyte, 2016).

Refrigeration is another field that superconductors are poised to revolutionize. While the concept of magnetically powered fridges has been around almost as long as standard gas-compression fridges, the technology to make these fridges viable has not been available until recently (Irving, 2016). Magnetically powered fridges have many advantages over traditional ones, such as requiring half the energy to run, and producing less noise and vibration. They are also simpler and more reliable in construction, requiring less maintenance work to keep them running (Irving, 2016). The first commercially available magnetic refrigerators have been on sale since 2016. These are large devices suitable for mass food and drink storage for businesses. Work is currently being done to make these fridges smaller and less expensive in order to make them valid options for everyday people to use in their homes (Irving, 2016).

Computers also stand to benefit from superconducting technology. Single flux quantum devices are computers which use superconductors and demonstrate ultra-high-speed processing combined with low power consumption (Kitazawa, 2011). While still under research, quantum computing is a growing field, and superconductors are essential to providing an efficient, high-power, low-heat environment for these computers to operate at their maximum potential (Kitazawa, 2011). Current consumer computers utilize heat sinks and fans in order to prevent overheating. This heat is generated by the internal components of the computer in the course of its operation as a waste product of the electricity being used. Using superconducting materials to construct computers would result in no heat generated as a waste product, removing the need for heat sinks and fans. This would allow computers to become much more compact and fit in more processing power in a smaller space. It would also make them far quieter and would extend the life of batteries as well.

Conclusion

Worldwide energy consumption will continue to increase as human civilization progresses. More people will gain access to higher standards of living and will need to be supported by an adequate supply of energy. The implementation of superconductors into numerous aspects of life will allow us to use energy more efficiently, reducing the overall amount we need to generate to maintain our society. Using superconductors will provide us with more land to use and allow us to transport and share energy across the world using entirely renewable sources. They will make renewable energy more feasible and reduce the hazards associated with massive amounts of electricity being transported through inhabited areas.

Transportation stands to be revolutionized through aspects of superconducting, such as magnetic levitation. Our food storage will become more efficient and reliable, and our computers will become smaller and more powerful than ever before. In the coming years, an increased ability to harness superconductivity will propel our global civilization to ever higher standards of living and sustainability.

References

Grant, P. M. (1997). Superconductivity and electric power: promises, promises ... past, present and future. IEEE Transactions on Applied Superconductivity, 7(2), 112–133. https://doi.org/10.1109/77.614432

Irving, M. (2016, June 16). Magnetic fridge eliminates gases, drastically reduces energy use. New Atlas. https://newatlas.com/cooltech-commercial-magnetic-cooling/43874/.

Islam, M. R., Guo, Y., & Zhu, J. (2014). A review of offshore wind turbine nacelle: Technical challenges, and research and developmental trends. Renewable and Sustainable Energy Reviews, 33, 161–176. https://doi.org/10.1016/j.rser.2014.01.085

Kitazawa, K. (2011). Superconductivity: 100th Anniversary of Its Discovery and Its Future. Japanese Journal of Applied Physics, 51(1). https://doi.org/10.1143/jjap.51.010001

Mikheenko, P., & Johansen, T. H. (2014). Smart Superconducting Grid. Energy Procedia, 58, 73–78. https://doi.org/10.1016/j.egypro.2014.10.411

New Health Advisor. (2020, May 15). Is Living near Power Lines Bad and Why? New Health Advisor. https://www.newhealthadvisor.org/Living-near-Power-Lines.html.

Ritchie,H. (2020). Energy Production and Consumption. Our World in Data. https://ourworldindata.org/energy-production-consumption.

Thomas, H., Marian, A., Chervyakov, A., Stückrad, S., Salmieri, D., & Rubbia, C. (2016). Superconducting transmission lines – Sustainable electric energy transfer with higher public acceptance? Renewable and Sustainable Energy Reviews, 55, 59–72. https://doi.org/10.1016/j.rser.2015.10.041

Whyte, C. (2016, June 14). How Maglev Works. U.S. Department of Energy. https://www.energy.gov/articles/how-maglev-works.

Wirfs-Brock, J. (2015, June 14). Lost In Transmission: How Much Electricity Disappears Between A Power Plant And Your Plug? Inside Energy. http://insideenergy.org/2015/11/06/lost-in-transmission-how-much-electricity-disappears-between-a-power-plant-and-your-plug/.

Conclusion

Superconductors were considered revolutionary when first discovered, and are beginning to demonstrate their potential impact on science and society. While many questions about their practicality remain, their wide range of potential applications, from transportation and energy to quantum computing and space travel, cannot be understated. With new and exciting developments in this field occurring all the time, the future of this technology is truly limitless and could bring about fundamental changes to our lives.

www.ingramcontent.com/pod-product-compliance
Lightning Source LLC
Chambersburg PA
CBHW031813190326
41518CB00006B/321